tech talk

Elementary
Workbook

John Sydes • Brian Cross

Contents

Unit 1

1 Complete the conversation with the words in the list.

Welcome Thank *you* little too meet slowly speak meet nice

A Excuse me. Are [1] ....*you*.... Ali Shaheen?
B Yes, I am. Nice to [2].................... you.
A Nice to [3].................... you, [4].................... . I'm Nikolay Aleksandrov.
[5].................... to Moscow.
B Thanks. It's [6].................... to be here.
A Here's your visitor's pass.
B [7].................... you.
A Can you [8].................... Russian?
B No, I can't. And I can only speak a [9].................... English.
A OK, no problem. I'll speak [10].................... .

2 Match a question or sentence on the left with a reply on the right.

1 *Hello. Are you Kiyoung Cho?*
2 Nice to meet you.
3 Welcome to Santa Barbara.
4 Can you speak Spanish?
5 What's this?
6 What's your extension?
7 I can only speak a little English.
8 Thanks.

a It's a map of the city.
b OK, I'll speak slowly.
c Hello, it's nice to meet you, too.
d Thank you. It's nice to be here.
e You're welcome.
f *Yes, I am. Nice to meet you.*
g It's 8089.
h No, I'm sorry, I can't.

3 Write these words as numbers.

1 My extension at work is four, five, oh, nine. 4509
2 Your room number is five, four, seven. ....
3 The area code for Bern is zero, three, one. ....
4 The serial number is two, eight, slash, double six, eight. ....
5 My home number is two, eight, oh, nine, three, four, five. ....
6 The model number is five dash six, zero, two, slash, five. ....
7 My flight number is BA six, double two, three. ....
8 The country code for Japan is double oh, eight, one. ....
9 The fax number is four, three, seven, zero, eight, six. ....
10 Your user number is double four, slash nine, oh, eight, dash, six, five, one. ....

4 Write *a* or *an*.

1 *an* umbrella
2 .... cell phone
3 .... American passport
4 .... car key
5 .... ID card
6 .... flashlight
7 .... owner's manual
8 .... Spanish dictionary
9 .... city map
10 .... old newspaper

5 Rearrange the letters to make a word.

1 nophe *phone*
2 krow ....
3 koccl ....
4 malar ....
5 chort ....
6 clepin ....
7 kittec ....
8 maulan ....
9 chanmie ....
10 beratty ....

6 **Complete the puzzle with the items in the pictures.**

1 b	a	t	t	e	r	y						
				2								
					3							
				4								
				5								
					6							
				7								
		8					■					

7 **Match a word on the left with a word on the right to make a phrase.**

1 *an English*	a code
2 a serial	b pass
3 an area	c clock
4 a visitor's	d phone
5 an identity	e number
6 a mobile	f manual
7 a user's	g *dictionary*
8 an alarm	h card
9 a city	i key
10 a door	j map

8 **Read the conversations and number the sentences in the correct order.**

Conversation 1

- [1] *Excuse me. Are you Fred Browning?*
- [3] *It's nice to meet you, too. I'm Greg Schultz.*
- [2] *Yes, I am. Nice to meet you.*

Conversation 2

- [1] Hello, I'm Paolo Castillo.
- [ ] Thank you.
- [ ] It's nice to meet you, too. Welcome to Electronika.
- [ ] Hello, I'm Dimitris Krassakis. Nice to meet you.

Conversation 3

- [ ] Yes, just a little.
- [1] Can you speak English?
- [ ] OK, no problem. I'll speak slowly.
- [ ] Thanks.

Conversation 4

- [ ] No, I'm sorry.
- [ ] Me too. So please speak slowly.
- [1] Can you speak French?
- [ ] I only speak a little English.
- [ ] OK. No problem.

Unit 2

1 Say the letters in each group. Circle the letter with a different sound.

1 b d t (q) 4 v w u q 7 r a j k
2 k a e j 5 s h m f 8 s x y m
3 b v g i 6 t i p e 9 c a e t

2 Which letters have the same sound as these words?

1 pay *a, j, k* 3 my .... 5 no ....
2 tea .... 4 day .... 6 car ....

3 Put the letters of the alphabet into the correct column.

day /eɪ/	**see** /iː/	**sell** /e/	**buy** /aɪ/	**go** /əʊ/	**you** /uː/	**are** /ɑː/
a	*b*	*f*	*i*	*o*	*q*	*r*

4 Match the numbers with the words.

35 ***72*** 49 **16** 28 ***60*** 81 **93** 57 ***27*** 82 **39** 13

sixteen .....
thirty-nine .....
forty-nine .....
twenty-eight .....
twenty-seven .....
thirty-five *35*
thirteen .....
seventy-two .....
eighty-one .....
eighty-two .....
fifty-seven .....
sixty .....
ninety-three .....

5 Write the next number in the sequence.

1 two, four, six, eight, ten, *twelve*
2 one, four, nine, sixteen, twenty-five, thirty-six, ...
3 three, five, seven, eleven, thirteen, seventeen, ...
4 five, eight, thirteen, twenty-one, thirty-four, fifty-five, ...
5 twelve, thirteen, twenty-four, fourteen, thirty-six, fifteen, ...
6 thirteen, twelve, fifteen, ten, seventeen, eight, ...

6 Complete the conversation using the phrases in the list.

How do you spell that	Yes, of course	what's your name
my last name's Maier	Thank you	*Can I help you*
I work for BMW	You're welcome	

A Good afternoon. [1] *Can I help you*?
B Good afternoon. Yes, I'm here to see Eduardo Silva.
A OK, and [2].., sir?
B It's Heiko Maier.
A [3]..?
B Heiko, that's H-E-I-K-O and [4]......................................,
that's M-A-I-E-R.
A Which company do you work for, Mr Maier?
B [5]..
A OK. Can you sign the form here, please?
B [6]..
A Just a second. I'll phone Mr Silva and tell him
you're here.
B [7]..
A [8].................................. . Please take a seat over there.
B Thanks.

7 Match the questions and sentences in column A with the answers in column B.

A	B
1 *How do you spell that?*	a Yes, ready.
2 Sorry, is that B or V?	b I work for BT.
3 Which company are you with?	c It's Maria Angelova.
4 How many do you need?	d Yes, that's all.
5 Are you ready?	e *It's G-R-E-S-H-A-M.*
6 What's your name?	f Yes, that's right.
7 What's the part number?	g You're welcome.
8 Thank you.	h That's B for Bravo.
9 So, that's 15 S-hooks.	i It's QBR slash two, six, one.
10 Is that it?	j Twenty.

8 **Find these eight items in the word puzzle.**

X	Y	W	A	S	H	E	R	S	Z
S	J	X	Z	Y	O	X	Q	Y	B
Q	P	J	X	Z	O	Y	B	J	A
X	Q	R	J	Z	K	Q	O	Z	T
Z	C	L	I	P	X	J	L	X	T
J	X	Z	Q	N	Y	X	T	Z	E
W	O	Q	Z	X	G	R	X	J	R
C	L	A	M	P	X	X	Y	Q	I
H	T	W	V	Y	Z	Q	Z	X	E
J	D	X	R	O	P	E	X	V	S

9 **Look at the visiting cards and complete the sentences.**

CHOU JUNG IH
After-Sales Service Engineer
Chi & Co Limited
C&C
3rd Fl., No.20,
Sec. 5 Roosevelt Rd.
Taipei
Taiwan
tel: 886-22-358-41424
fax: 886-22-358-06420

ABR
Kurt Brausing
Service Technician
ABR Danmark
Helgeshoej Allé 26
DK - 2630
Taastrup
Denmark
tel: 45-43-410900
fax: 45-43-410080

EnTech
Joey Yap Souza
Technical Staff Software Engineer
EnTech Equipamentos Corporation Ltd
90 Thomson Road,#25-06 United Square,
307591Singapore
tel: 65-254-66760
fax: 65-254-66425

East African Technologies Limited.
Abdel Ashraf
Maintenance
P.O.Box 46650, Waiyaki Way,
ABC Place, Nairobi
Kenya
tel: 254-2-441-4949
fax: 254-2-448-5000

PAOLA BELISARIO
Quality Assurance Manager
ATOL (ITALIA) S.p.A.
Via Mario Pannuzio 3,
20156 Milano, Italy
tel: 39-02-280 9235
fax: 39-02-280 9236

Equipamentos Indústrias Ltda
Taís de Souza Partelli
Production Planner
Rua Sumidouro 90
05428-7751
São Paulo
Brazil
tel:55-11-3211-56120
fax:55-11-3211-56222

1 The after-sales service engineer works for ....
2 Abdel Ashraf works for ....
3 The Brazilian production planner works for ....
4 Kurt Brausing is a .... and works for ....
5 Joey Yap Souza works in .... for .... He's a ....
6 The Italian quality assurance manager's name is ....

10 **What about you? Write three sentences about yourself.**

1 I work in ....
....
2 I work for ....
....
3 I am a / an ....
....

Unit 3

1 Match the pictures with the items in the menu.

1 .... 4 .... 7 ....

2 .... 5 .... 8 ....

3 .... 6 .... 9 ....

Menu

Cola / Diet Cola	small	**1.20**	medium	**1.50**	large	**1.75**
Coffee	small	**1.15**	medium	**1.40**	large	**1.65**
Milkshake	small	**1.35**	medium	**1.55**	large	**1.80**
Burger	**2.15**					
Cheeseburger	**2.30**					
Hot dog	**1.95**					
Sandwich	ham		**2.50**			
	chicken		**2.75**			
	roast beef		**2.95**			
French fries	small	**1.00**	medium	**1.25**	large	**1.50**
Doughnut	**1.80**					

2 Complete the conversation with the words and phrases in the list.

And something to drink	Can I have	Thanks.	How much is it
Can I help you	a large	*me, too*	Large, medium, or small
Have a nice day.	please		

TOM Let's get something to eat. I'm very hungry.

ANDREAS Yes, [1] *me, too* .

SERVER [2].. ?

TOM Yes, a Diet Cola and a cheeseburger, please.

SERVER A Diet Cola. [3].................................. ?

TOM Medium, please. And one large French fries, [4].................................. .

SERVER And for you, sir?

ANDREAS A hot dog and [5].................................. French fries, please.

SERVER [6].. ?

ANDREAS Yes, a large shake, please. And a doughnut.

TOM [7].................................. a doughnut too, please?

SERVER Yes, no problem.

TOM [8].................................. .

SERVER Coming right up. Here you are.

TOM Great! Thanks. [9].. ?

SERVER That's fourteen euros and fifteen cents.

TOM Here you are.

SERVER Thanks. [10].................................. .

3 Match the sentences and questions with the correct answer.

1 *I'm hungry.* — e
2 Can I help you?
3 Do you want a doughnut?
4 Have a nice day.
5 How much is it?
6 Let's get something to eat.
7 Large, medium, or small?

a Yes, a large coffee, please.
b Three euros forty, please.
c Good idea!
d No, thanks.
e *Me, too.*
f Medium, please.
g Thanks.

4 Write three more conversations. Follow the example.

have / some paper — *Can I have some paper*?
write down the reference number? — *Do you want to write down the reference number*?
make a list of parts — *No, I want to make a list of parts.*

1 use / your mobile phone ....?
make a call? ....?
send a text message ....

2 have / the order form ....?
change the order? ....?
check the quantity ....

3 look at / your manual ....?
read the instructions? ....?
check a serial number ....

5 Tick (✓) the sentences that are correct. Correct the sentences that are wrong.

1 *It's a six-metre cable.* — ✓
2 *This is a 30-centimetre~~s~~ ruler.* — *a 30-centimetre*
3 I need a 10-amp fuse. ....
4 Do you want a 60 or 75-watts bulb? ....
5 It has a 1.6-litre diesel engine. ....
6 I want two three-inch screws. ....
7 How many six-volts batteries do you need? ....
8 Is it a five-litres tank? ....

6 Complete the list of measurements.

1 1,000 m = *1 km*
2 3 ft = ...
3 100 cm = ...
4 8 pints = ...
5 16 oz = ...
6 10 mm = ...
7 12 ins = ...
8 1,000 mg = ...

7 Write the measurements in the correct column.

Length	Volume	Weight
kilometres	*pints*	*pounds*

kilometres	millilitres	centimetres
gallons	tonnes	litres
yards	*pounds*	kilograms
milligrams	metres	millimetres
feet	inches	*pints*
ounces	grams	

8 Karl needs some parts for his motorbike. Complete the conversation with the words and phrases in the list.

I need	How much is it	Is it	Do you have	want
this bolt	Do you want	no good	it is	I need a

ASSISTANT Can I help you?

KARL Yes, [1] *I need* a new battery for my motorbike.

ASSISTANT OK, do you [2] .... this one?

KARL Is it twelve volts?

ASSISTANT No, it isn't. It's six volts.

KARL That's [3] .... I need a twelve-volt battery.

ASSISTANT OK. Anything else?

KARL Yes, a cable.

ASSISTANT [4] .... the 900 mm cable or the 1500 mm cable?

KARL The 900 mm cable. And I need a long bolt.

ASSISTANT OK. Do you want [5] ....?

KARL [6] .... 50 mm?

ASSISTANT Yes, [7] ....

KARL That's no good. [8] .... 60 mm bolt. And I need a spring too. [9] .... a 10 mm spring?

ASSISTANT Yes. Here you are.

KARL Great, thanks. [10] ....?

ASSISTANT It's $17.50.

9 Read the conversation in 8 again. Tick (✓) the items on the order form that Karl needs.

Parts order form

☐ 10 mm spring	☐ 6 volt battery	☐ 50 mm bolt	☐ 900 mm cable
☐ 12 mm spring	☐ 12 volt battery	☐ 60 mm bolt	☐ 1500 mm cable

Unit 4

1 Complete the conversation. Underline the correct phrase.

KARIN Good morning, Butler Ltd. [1]*Can I help you?* / *Who are you?*
CATHERINE Good morning. [2]*Can I* / *Do I* speak to Max Schwarz, [3]*thanks* / *please?*
KARIN [4]*I'm not* / *I'm afraid* he's not here today.
CATHERINE Oh, can you [5]*give* / *say* me Max's mobile number?
KARIN No, I'm afraid I [6]*can't* / *don't*. It's a personal number.
CATHERINE OK. Can you [7]*take* / *leave* a message, please?
KARIN No problem. [8]*Just* / *Only* a second, please. I [9]*need* / *find* a pen. OK. What's your name, please?
CATHERINE [10]*Here is* / *It's* Catherine Maubeuge.
KARIN Maubeuge? How do you [11]*say* / *spell* that, please?
CATHERINE M-A-U-B-E-U-G-E. Please [12]*ask* / *say* Max to [13]*speak* / *call* me. My phone number is 00 33, that's the code for France, 128 3900.
KARIN So, [14]*that's* / *you're* 00 33 128 3900.
CATHERINE That's [15]*OK* / *correct*.
KARIN And what company are you [16]*with* / *for*, Ms Maubeuge?
CATHERINE Renault.
KARIN OK.
CATHERINE Thank you very much.
KARIN [17]*Please* / *You're welcome*. Have a nice day.

2 Read the messages. Which one is correct?

2 Max,
I want to speak to
Catherine Maubeuge
from Renault.
Phone: 00 33 128 3900

3 Match the sentences with the correct reply.

1 *Can I speak to Paulo, please?*
2 Can you take a message?
3 So that's 0897 6628 0923.
4 What's the code for Munich?
5 Are you ready?
6 How do you spell that?
7 This is Laura Bray.
8 His name's Lettl.
9 Goodbye.

a Yes, ready.
b Yes, that's right.
c Hello, Laura. How are you?
d *I'm afraid he's not here.*
e How do you spell that?
f Sure. Just a second. I need a pen.
g Bye.
h It's 089.
i S-O-U-S-A.

4 Look at the equipment list. You need some of these items. Complete the conversation.

Doc. Ref: Drb 04-11

Date : 03 April

Stock control Storeroom D: Peter BERNIE / electrical

ref.	item	quantity in stock
203/11-9	10 x 9 V battery	24
203/10-1.5	16 x 1.5 V battery	40
199/07-25	10 x 25 W bulb	22
199/07-40	20 x 40 W bulb	10
361/22-10	20 x 10 amp fuse	53
361/22-13	40 x 13 amp fuse	25
524/04-1x100	4 x 1 mm cable (100m)	12
524/05-1.5x50	2 x 1.5 mm cable (50m)	14
524/06-1.5x100	2 x 1.5 mm cable (100m)	10

YOU I [1] ....*need*.... twenty 9-volt batteries.
PETER No problem, we [2] ....*have*.... twenty-four.
YOU And I [3] .... ten 25-watt bulbs.
PETER OK. I [4] .... twenty-two. No problem.
YOU [5] .... have any 40-watt bulbs?
PETER Yes. How many [6] .... need?
YOU Twenty.
PETER Sorry. I only [7] .... ten.
YOU That's no good. What about 60-watt bulbs?
PETER No, I [8] .... any 60-watt bulbs, only 25-watt or 40-watt.
YOU Do [9] .... any fuses?
PETER Yes, what fuses [10] .... need?
YOU I need twenty 10-amp fuses, and forty 13-amp.
PETER OK. I [11] .... enough 10-amp fuses, but I [12] .... enough 13-amp fuses.

5 Rearrange the letters to make the names of these items.

1 llaw cokest ....*wall socket*....
2 neonexits deal ....
3 sinbuses darc ....
4 sailer rubmen ....
5 nationallist kids ....
6 hotelepen burnme ....
7 amile sadders ....

6 Follow the instructions to draw a picture.

1 Draw a diagonal line from sixty-six to fifty-seven.
2 Draw a horizontal line from eighty-four to eighty-seven.
3 Draw a vertical line from fourteen to sixty-four.
4 Draw three horizontal lines, one from thirteen to fourteen, one from forty to forty-one, and one from ninety to ninety-one.
5 Draw two horizontal lines, one from fourteen to nineteen and one from sixty-four to sixty-nine.
6 Draw three vertical lines, one from twenty-eight to sixty-eight, one from sixty-six to twenty-six, and one from twenty-seven to fifty-seven.
7 Draw three diagonal lines, one from forty to thirteen, one from fourteen to forty-one, and one from ninety-one to sixty-four.
8 Draw three vertical lines, one from nineteen to sixty-nine, one from forty to ninety, and one from ninety-one to forty-one.
9 Draw two horizontal lines, one from seventy-eight to seventy-five and one from twenty-six to twenty-eight.
10 Draw two more diagonal lines, one from eighty-four to seventy-five and one from seventy-eight to eighty-seven.

10	11	12	13	14	15	16	17	18	19
20	21	22	23	24	25	26	27	28	29
30	31	32	33	34	35	36	37	38	39
40	41	42	43	44	45	46	47	48	49
50	51	52	53	54	55	56	57	58	59
60	61	62	63	64	65	66	67	68	69
70	71	72	73	74	75	76	77	78	79
80	81	82	83	84	85	86	87	88	89
90	91	92	93	94	95	96	97	98	99

7 Choose the correct answer.

Can I speak to John, please?
a No, he's not.
b *Sure. Just a second.*
c Coming right up.

Can I leave a message?
a No, thanks.
b You're right!
c Yes, no problem.

Can you spell that, please?
a Yes, it's 0041 225 6587.
b So that's H-Y-W-E-L.
c Yes, of course. Frenwich, that's F-R-E-N-W-I-C-H.

Can you call me later?
a Yes, no problem.
b Just a second.
c Yes, that's right.

Can I speak to Anthony Boyle, please?
a Yes, this is Anthony Boyle speaking.
b I am Anthony Boyle.
c Yes, it's me.

My telephone number is 0047 21 66405.
a That's right.
b So, that's 0047 21 66405.
c No, 66415.

I'm afraid David isn't here. Can ...
a I take a message?
b you take a message?
c you call me about this?

So, you need three user guides and two installation disks?
a Yes, I need them.
b Yes, you're correct.
c Yes, that's right.

My number at work is 0171, that's the area code, 441489 and ...
a my extension is 3184.
b my address is 3184.
c my department is 3184.

Thank you for calling.
a Thank you. Bye.
b You're welcome. Bye.
c Please. Bye.

Unit 5

1 Match the controls in the diagram with the words in the list.

1 levers ..*H*...
2 LCD displays .....
3 knobs .....
4 switches .....
5 clock .....
6 gauges .....
7 sockets .....
8 key pad .....

2 Look at the diagram and complete the sentences with *There's* or *There are*.

1 ....*There's*.......... a gauge in the centre on the left.
2 .... two levers on the right at the bottom.
3 .... four switches at the bottom in the centre.
4 .... a clock on the left at the top.
5 .... an LCD display in the centre.
6 .... three gauges on the right at the top.
7 .... a socket on the left in the centre.
8 .... a key pad in the centre.

3 Look at the keypad and the diagrams. Write the number of the key next to its position.

1 at the top, on the right ..*3*...
2 in the centre, above the zero .....
3 on the left, below the four .....
4 at the top, in the centre .....
5 at the bottom, on the left .....
6 on the right, below the nine .....

4 Read the instructions. Draw the controls on the diagram.

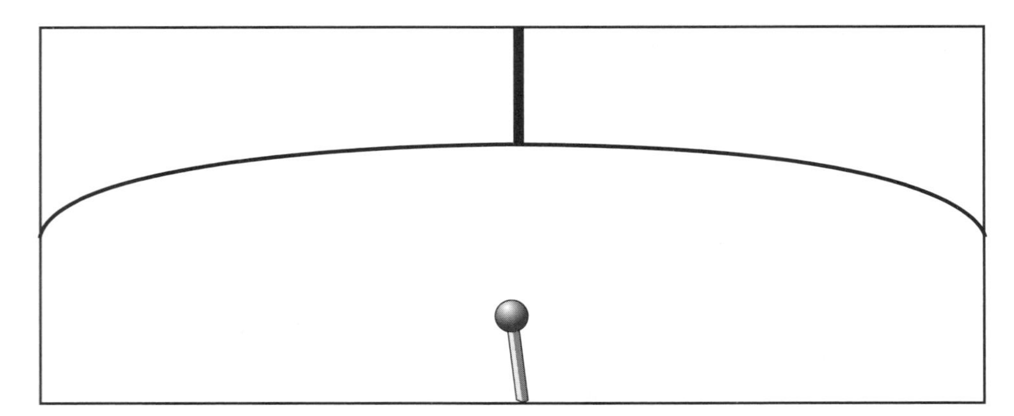

1 The lever is in the centre at the bottom.
2 There are three sockets in the middle on the right.
3 There are two knobs at the top on the left.
4 There's a key pad in the centre.
5 There are three gauges on the left in the centre.
6 The clock is on the right at the top.
7 There are two switches on the left at the bottom.
8 The LCD display is in the middle at the top.
9 The air vent is on the right at the bottom.

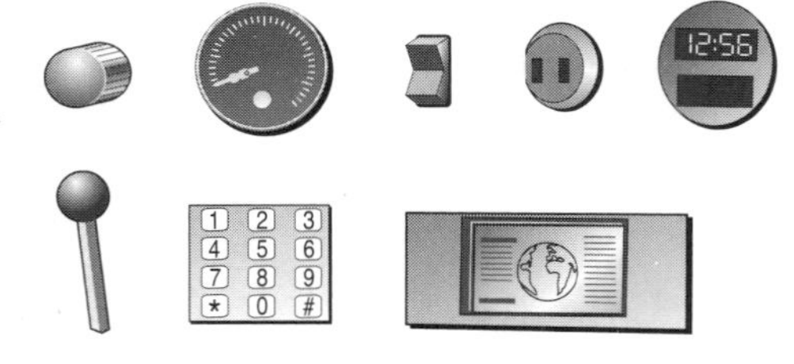

5 Look at the alarm control panel and complete the instructions. Use the words in the list.

press	*open*	button	small	correct	lock
top	centre	right	replace	switches	display

OK, are you ready? Turn the key at the bottom and [1] *open* the cover.

There's a clock at the [2].................. . Look at the clock and check the time is [3].................. . There's also a [4].................. button at the top, on the left. [5].................. the button. This [6].................. on the alarm.

Now, there's a key pad in the [7].................. . Type in the code on the key pad. The code is 65786.

Now, look at the LCD [8].................. under the clock. If it says 'Ready', press the large [9].................. at the bottom, on the [10].................. . Close the cover, and [11].................. it with the key.

That's it! Remember to [12].................. the batteries!

6 Write the opposites.

1 lock .... *unlock*
2 open ....
3 replace ....
4 turn off ....
5 push ....
6 turn anti-clockwise ....
7 unplug ....

7 Underline the correct instruction.

1 *Open* / *Unplug* / *Pull* the air-vent.
2 *Read* / *Unlock* / *Turn* the LCD display.
3 *Lock* / *Turn* / *Read* the knob anti-clockwise.
4 *Switch on* / *Pull* / *Close* the lever.
5 *Plug in* / *Switch off* / *Unlock* the door.
6 *Lock* / *Open* / *Replace* the bulb.
7 *Press* / *Turn* / *Open* the button.
8 *Remove* / *Close* / *Lock* the fuse.

8 Complete the instructions with the correct words.

1 .....*Open*..... the cover, ...*replace*..... the batteries, and .....*close*..... the cover.
(*close, open, replace*)

2 .... your PC, ... your email program, and .... your new messages.
(open, switch on, read)

3 When you go, please .... the windows, ... the lights, and .... the door.
(close, lock, switch off)

4 .... the electricity, ... the 60-watt bulb, and .... it with a 75-watt bulb.
(remove, replace, switch off)

5 To stop the machine, ... the lever, ... the knob on the right anti-clockwise, and .... the motor with the on / off switch at the bottom on the left.
(push, turn, turn off)

9 Write the plurals of these words.

1 box .... *boxes*
2 car ....
3 business ....
4 battery ....
5 person ....
6 day ....
7 bulb ....
8 torch ....
9 city ....
10 month ....
11 item ....
12 message ....

10 **Read the conversations and mark the hotel facilities on the diagram.**

1 A Good morning, madam. Can I help you?
 B Yes, where's the café?
 A It's on the second floor.
 B Thanks.

2 A Excuse me, where's the fitness centre?
 B It's on the sixth floor.
 A Is there an elevator?
 B Yes, the elevator is on the left next to the stairs.

3 A Where's the bar?
 B Well, there's a cocktail bar on the seventh floor on the left, next to the Mexican restaurant. And there's a bar on the second floor next to the café.
 A Thanks. And is there a photocopier here?
 B Yes, sir. There are two photocopiers on the fourth floor next to the meeting room.

4 A Excuse me, where can I park my car?
 B The hotel car park is on the right.
 A OK. Is there a swimming pool?
 B Yes, sir. The swimming pool and sauna are on the lower level.

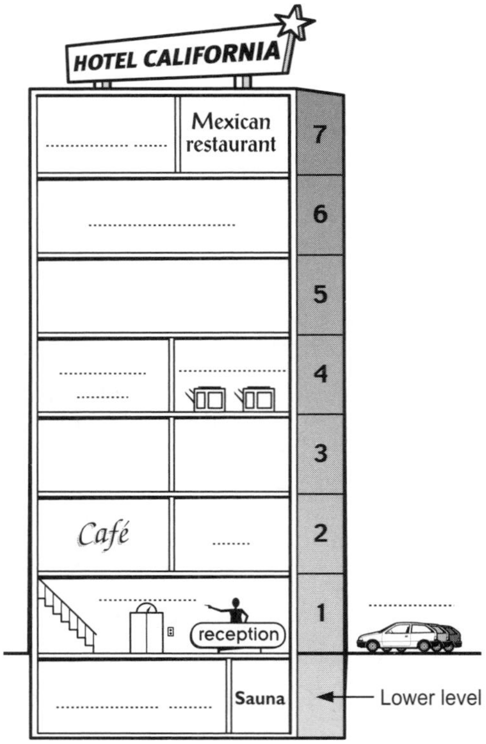

HOTEL CALIFORNIA
Mexican restaurant
7
6
5
4
3
Café
2
1
reception
Sauna
Lower level

11 **Read the conversations again. Answer the questions using** ***Yes, there is / Yes, there are*** **or** ***No, there isn't / No, there aren't.***

1 Is there a fitness centre on the sixth floor?
 Yes, there is.

2 Are there any bars in the hotel?
 ....

3 Is there a café on the third floor?
 ....

4 Is there a car park on the lower level?
 ....

5 Are there any photocopiers on the fourth floor?
 ....

6 Are there any meeting rooms on the second floor?
 ....

7 Is there a cocktail bar on the first floor?
 ....

8 Are there any elevators?
 ....

Unit 6

1 Match the parts of the car with the words in the list.

1	driver's seat	E.....	7	bonnet (BrE) / hood (AmE)	.....
2	steering wheel	.....	8	boot (BrE) / trunk (AmE)	.....
3	gear lever	.....	9	sunroof	.....
4	windscreen (BrE) / windshield (AmE)	.....	10	CD player	.....
5	headlights	.....	11	navigation system	.....
6	aerial (BrE) / antenna (AmE)	.....	12	aluminium wheels	.....

2 Read the order form. Complete the conversation on page 20 with the phrases in the list.

Yes, it does   Do you want   No, it doesn't   Does it have

Order form

Customer name *David White*   Salesperson *Philip Bray*
Model *Mini Cooper*   Colour *Green*

Standard features

- ☐ **Metallic paint**
- ☐ **Light metal wheels**
- ☐ **Air bags** (driver, passenger, side)
- ☐ **Radio**
- ☐ **Cruise control**

Extras

- ☐ **CD player**
- ☐ **Air-conditioning**
- ☐ **Sunroof**
- ☐ **Automatic transmission**
- ☐ **Telephone**
- ☐ **Electrically-powered seats**
- ☐ **Heated locks**
- ☐ **Head air bags**
- ☐ **Leather seats**
- ☐ **Navigation system**
- ☐ **Alarm**
- ☐ **Central locking**

DAVID	OK, I want a metallic green Mini Cooper with light metal wheels, and air bags.
SALESMAN	The air bags are standard. [1] *Do you want* head air bags too?
DAVID	Yes, please. [2] .... a sunroof as standard?
SALESMAN	[3] .... It's £430 extra.
DAVID	£430, hmm. OK.
SALESMAN	Do you want a sunroof?
DAVID	Yes, I want a sunroof. [4] .... a navigation system?
SALESMAN	[5] .... The navigation system needs a radio to work, but the radio is standard on all models.
DAVID	OK. And I want a CD player too.
SALESMAN	What about heated locks?
DAVID	No, I don't need them. [6] .... central locking?
SALESMAN	[7] .... So, central locking too?
DAVID	Yes, I need central locking. And what about automatic transmission?
SALESMAN	Automatic transmission isn't standard. [8] .... that too?
DAVID	Yes, I do.

3 Complete the order form in 2. Tick (✓) the extras the customer wants.

4 Mark the sentences true (T) or false (F).

1 A map is made of paper. (T)/F
2 Ropes are made of nylon. T/F
3 A book is made of aluminium. T/F
4 Car tyres are made of rubber. T/F
5 Jogging shoes are made of wood. T/F
6 A box is made of cardboard. T/F
7 Sweaters are made of wool. T/F
8 A table is made of ceramic. T/F

5 Write what these things are made of. Use *It's made of* (1 thing) or *They're made of* (2 things or more).

1 packing material *It's made of polystyrene.*
2 hooks *They're made of steel or iron.*
3 a window ....
4 a door ....
5 the cover of a switch ....
6 coffee filters ....
7 a car engine ....
8 shoes ....
9 elevator cables ....
10 a dictionary ....

6 Answer the questions with *Yes, it does* or *No, it doesn't*.

1 Does a car have four wheels? *Yes, it does.*
2 Does a motorbike have several air bags? ....
3 Does a telephone have a lot of levers? ....
4 Does a fuse box have several switches? ....
5 Does a PC have a fan? ....
6 Does a cable have a lot of sockets? ....
7 Does a windscreen/windshield have a heater? ....
8 Does an elevator have an air vent? ....

7 Underline the correct answer.

1 A bus has a lot of *wheels* / *seats* / *engines*.
2 A computer has several *fans* / *sockets* / *levers*.
3 An electric car doesn't have a(n) *fuel tank* / *air bag* / *brake pedal*.
4 A mobile phone has a lot of *batteries* / *alarm clocks* / *buttons*.
5 A dictionary has a lot of *maps* / *words* / *news*.
6 A CD player has a *drive* / *minibar* / *elevator*.
7 A good hotel has a lot of *stairs* / *facilities* / *instructions*.
8 Internet addresses don't have an @ / _ / é.
9 A country code has several *numbers* / *IDs* / *letters*.
10 A torch has a *bulb* / *pump* / *fuse*.

8 Look at the shapes. Fill in the missing letters in the words. Write what the shape is.

1 It's a ci*r*c*l*e.
It's *round* or *circular*.

2 It's a __em__-ci__c__e.
It's ....

3 It's an __v__l .
It's ....

4 It's a c__l__n__er.
It's ....

5 It's a t__ __an__ __e.
It's ....

6 It's a __p__ er__ .
It's ....

7 It's a c__b__ .
It's ....

8 It's a __qu__r__ .
It's ....

9 What shape are these things?

1 a compact disk *It's round.*
2 an Olympic-sized swimming pool ....
3 a room that is 4 metres x 4 metres ....
4 one side of a pyramid ....
5 a fuel pipe ....
6 a half-moon ....
7 a table-tennis ball ....
8 an egg ....
9 an oil tank that is 1.5m x 1.5m x 1.5m ....

Unit 7

1 Look at AIBO, the robot dog. Label the parts of his body with the words in the list.

face	back
mouth	nose
neck	front leg
back leg	foot
head	

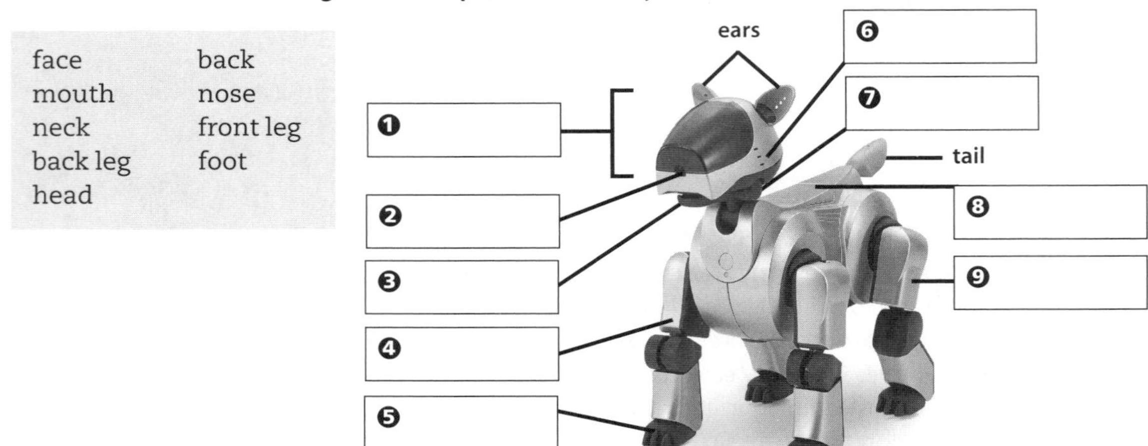

2 Read about AIBO, the robot dog. Mark the sentences true (T) or false (F).

AIBO (Artificial Intelligence roBOt) is a robot dog, developed by Sony. He is 27.4 cm long, 26.6 cm high, and weighs 1.5 kilos. You can control him with a computer, a remote control, or with voice instructions. AIBO is not just a toy. Sony developed and tested him as an electronic friend for old people.

Movable parts

- Mouth: 1 degree of freedom
- Head: 3 degrees of freedom
- Legs: 3 degrees of freedom x 4 legs (removable)
- Ears: 1 degree of freedom x 2 ears
- Tail: 2 degrees of freedom
- Total: 20 degrees of freedom

He needs a 9V battery and can operate for approximately 1.5 hours.

Price: approximately $1,300.

Here are some things AIBO can do:

- walk and run backwards and forwards
- stand on his back legs and raise and lower his front legs
- see and remember things
- understand the name you give him
- dance
- understand 75 different voice instructions (*Sit! Come! No!*, etc.)
- take photographs
- make robotic dog sounds
- learn new actions.

1 AIBO is a toy for children. T/(F)
2 He can remember his name. T/F
3 He can bend his legs. T/F
4 He can stand on his front legs. T/F
5 He can raise and lower his front legs. T/F
6 He can turn on a light. T/F
7 He can speak 75 words of English. T/F
8 He can learn to do new things. T/F
9 He can move his ears and tail. T/F
10 He can bend his back. T/F

3 **Match the pictures with the verbs.**

1 move backwards, move forwards *A*
2 open, close .....
3 push, pull .....
4 adjust .....
5 bend, straighten .....
6 turn left, turn right .....
7 climb up, climb down .....
8 carry .....
9 raise, lower .....

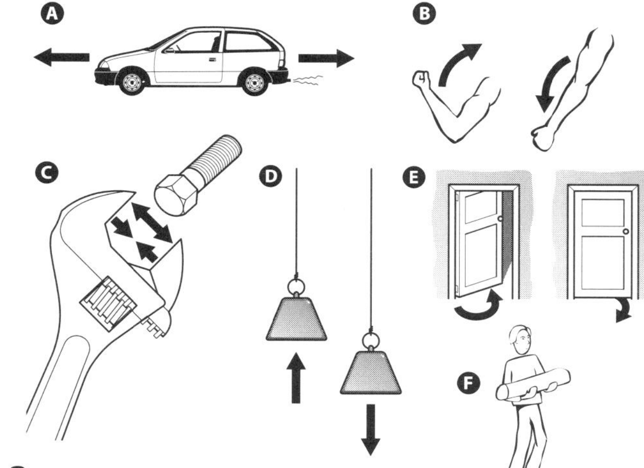

4 **Complete the sentences with a verb from 3.**

1 Please *close* the door and lock it when you go home.
2 A lot of robots can't .... up or down stairs.
3 Please help me .... the car into the garage.
4 Please .... your hand if you want to ask a question.
5 Can it .... backwards and forwards, or only forwards?
6 The maximum load this robot can .... is 25 kg.
7 ASIMO can .... the size of his steps to make them big or small.
8 When you want to lift something heavy, ... your legs.
9 Do I .... left or right when I come out of the elevator?
10 There are so many bends in this road – we need to .... it.

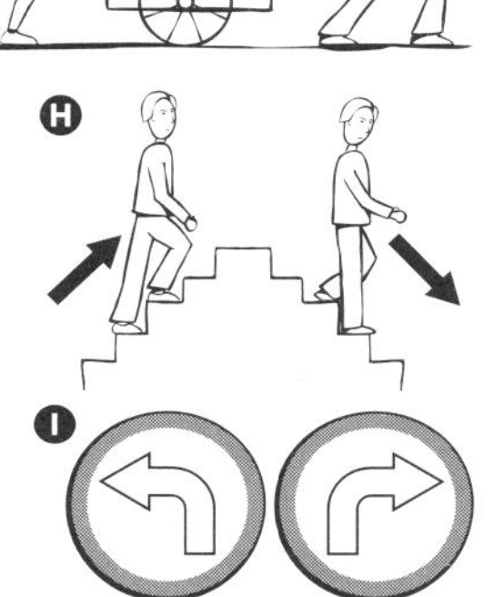

5 **What is it? Complete the sentences with the items in the list.**

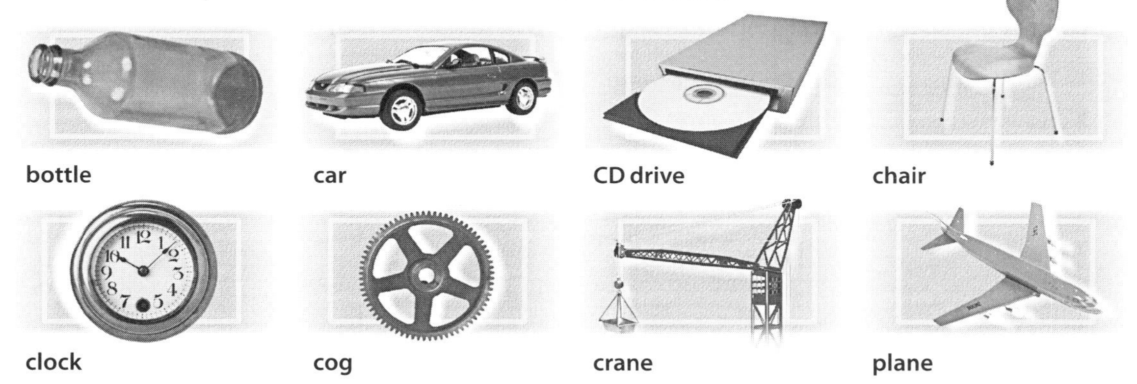

bottle **car** **CD drive** **chair**
clock **cog** **crane** **plane**

1 It has a neck and a mouth, but it can't drink. *bottle*
2 It has arms, legs, and a back, but it can't move. ....
3 It has two or three hands. It doesn't have a mouth, but it can tell you the time. ....
4 It's round. It has a lot of teeth, but it can't eat. ....
5 It has a nose. It can't smell, but it can fly. ....
6 It has an arm, but it doesn't have a hand. It can lift things. ....
7 It can read and write, but it doesn't have eyes or hands. ....
8 It has a body. It doesn't have arms or legs, but it can carry you somewhere. ....

6 Match the pictures with the dimensions.

A Solar-powered radio

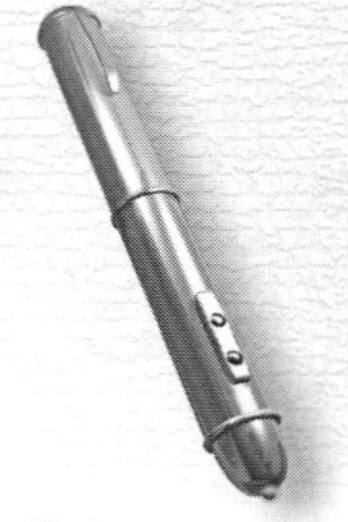

B Laser pointer

C Electronic dictionary

D Computer table

E Solar panel

F Electronic door opener

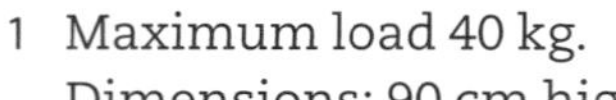

1 Maximum load 40 kg.
Dimensions: 90 cm high, 70 cm wide, 55 cm deep.
Price €99.99.

2 Operates on 12 volts.
Dimensions: 81 mm long, 60 mm high, 22 mm wide.
Price €49.99.

3 Dimensions: 150 mm long, 70 mm high, 30 mm wide.
Price €29.95.

4 25-year guarantee. Weight, 7.5 kilos.
Dimensions: 1,200 mm high, 500 mm wide, 60 mm deep.
Price €539.50.

5 Weight without batteries, 125 g.
Dimensions: 135 mm long, 90 mm wide, 15 mm high.
Price €134.99.

6 Wavelength 670 mm (red).
Dimensions: 15 cm long, 13 mm wide.
Price €29.95.

7 Look at the answers and complete the questions. Then write four more questions.

1	How *high* is the computer table?	90 cm
2	How .... is the solar panel?	500 mm
3	How .... is the laser pointer?	15 cm
4	How .... is the electronic dictionary?	125 g
5	How .... is the electronic door opener?	€49.99
6	How .... is the guarantee on the solar panel?	25 years
7	How .... is the electronic dictionary?	135 mm
8	How .... is the solar radio.	70 mm

8 Write the opposites.

1	up	*down*	7	forwards	....
2	....	dirty	8	....	turn right
3	large	....	9	bend	....
4	....	can't	10	....	pull
5	upstairs	....	11	turn on	....
6	....	raise	12	....	close

Unit 8

1 Read the dialogue and complete the task list.

A Can I help you?
B My car needs a service.
A OK. Any problems?
B Yes. The radio doesn't work. I need a new one.
A No problem. We'll install it. Anything else?
B Yes. I think I need four new tyres.
A OK, we'll check the tyres and replace them if you need new ones.
B Thanks.
A Is that all?
B No. I can't unlock the passenger door.
A Do you have central locking?
B No, I don't.
A No problem. We'll replace the old lock with a new one.
B Great.
A Do you need anything else?
B Yes. I need an alarm. Do you have one for this car?
A Yes, no problem. We'll install it for you.
B Great. Can you do it today?
A No, sorry. We're very busy.

Task list

Vehicle: Ford Focus XE53 BJP
Date: 17 August
1 .... a new radio.
2 .... and 3 .... the tyres.
4 .... the passenger door lock.
5 .... an alarm.

2 Complete the conversations with *it* or *them*.

1 A We need to check how long the room is.
B Right, I'll measure ......it...... .
2 A I need a new password.
B OK, I'll change .... for you.
3 A These bolts are loose.
B Do you want me to tighten ....?
4 A We need to send these parts today.
B OK, I'll get a box and pack ....
5 A The fire doors are locked.
B Are they? I'll unlock ....
6 A There's a problem with this machine.
B OK, we'll call maintenance and ask them to service ....
7 A Are these figures correct?
B I'm not sure. I'll check ....
8 A There's a hole in this hose.
B Yes, we need to replace ....

3 Complete the conversations with the verbs in the list.

carry replace clean check straighten measure order delete

1 A This box is very heavy.
  B OK, I'll help you *carry* it.
2 A We don't have enough paper.
  B No problem. I'll .... some more.
3 A This aerial is bent.
  B OK, I'll .... it.
4 A These figures can't be correct.
  B Are you sure? I'll .... them.
5 A Is this 25 cm long?
  B I'm not sure, but I'll .... it.
6 A This light bulb doesn't work.
  B I know. I'll .... it.
7 A The car's very dirty.
  B Yes, I'll .... it tomorrow.
8 A I don't need these programs.
  B OK, I'll .... them for you.

4 Underline the correct word.

1 A I need to measure this room.
  B Do you need [1]*a / some* tape measure?
  A Yes. Do you have [2]*one / any*?

2 A I need to straighten these wires.
  B Do you need [3]*a / some* pliers?
  A Yes. Do you have [4]*one / any*?

3 A I need to remove these screws.
  B Do you need [5]*a / some* screwdriver?
  A Yes. Do you have [6]*one / any*?

4 A I need to replace that light bulb.
  B Do you need [7]*a / some* ladder?
  A Yes. Do you have [8]*one / any*?

5 A I need to hang this picture on the wall.
  B Do you need [9]*a / some* hammer and [10]*a / some* nails?
  A I don't need [11]*a / some* hammer, but I need [12]*a / some* nails. Do you have [13]*one / any*?

5 **Read the FAQs 1–10 and match them with the answers a–j.**

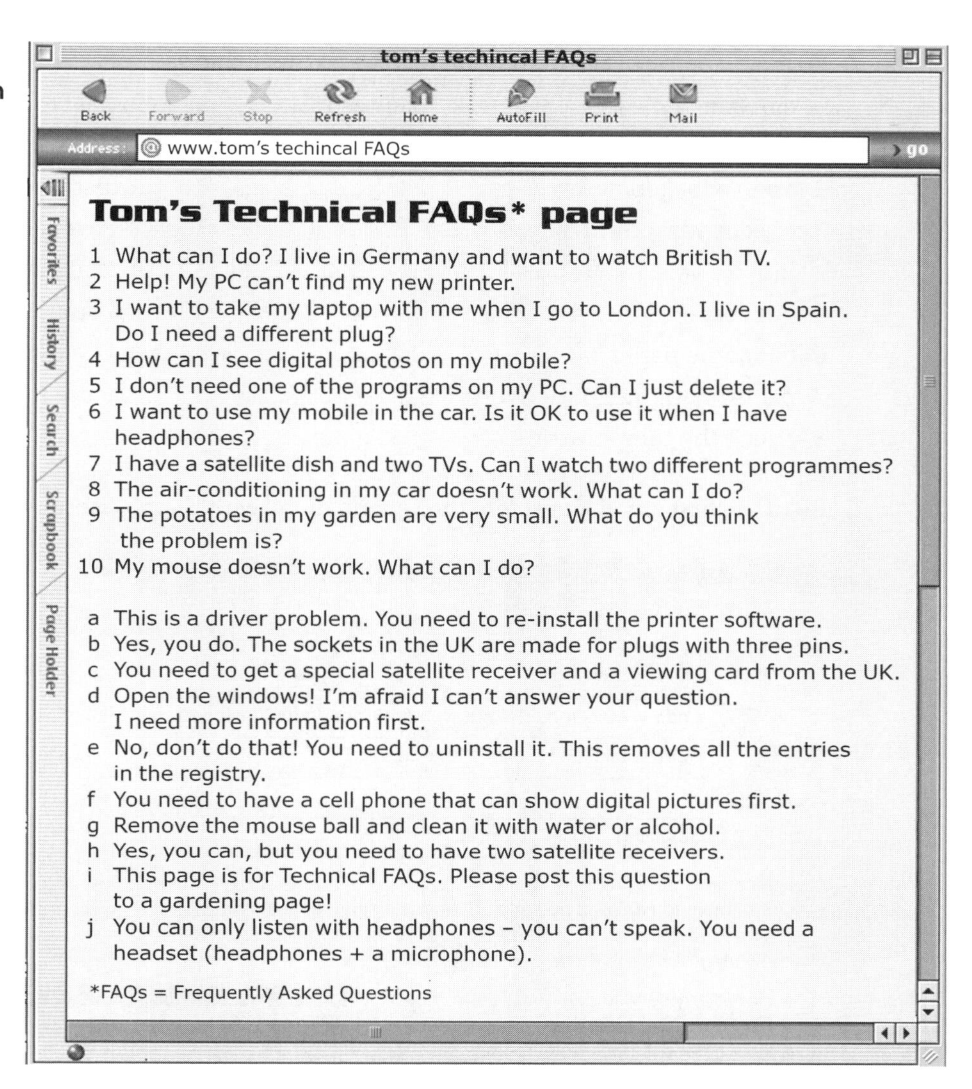

tom's techincal FAQs

Back Forward Stop Refresh Home AutoFill Print Mail

Address: www.tom's techincal FAQs

Favorites History Search Scrapbook Page Holder

Tom's Technical FAQs* page

1 What can I do? I live in Germany and want to watch British TV.
2 Help! My PC can't find my new printer.
3 I want to take my laptop with me when I go to London. I live in Spain. Do I need a different plug?
4 How can I see digital photos on my mobile?
5 I don't need one of the programs on my PC. Can I just delete it?
6 I want to use my mobile in the car. Is it OK to use it when I have headphones?
7 I have a satellite dish and two TVs. Can I watch two different programmes?
8 The air-conditioning in my car doesn't work. What can I do?
9 The potatoes in my garden are very small. What do you think the problem is?
10 My mouse doesn't work. What can I do?

a This is a driver problem. You need to re-install the printer software.
b Yes, you do. The sockets in the UK are made for plugs with three pins.
c You need to get a special satellite receiver and a viewing card from the UK.
d Open the windows! I'm afraid I can't answer your question. I need more information first.
e No, don't do that! You need to uninstall it. This removes all the entries in the registry.
f You need to have a cell phone that can show digital pictures first.
g Remove the mouse ball and clean it with water or alcohol.
h Yes, you can, but you need to have two satellite receivers.
i This page is for Technical FAQs. Please post this question to a gardening page!
j You can only listen with headphones – you can't speak. You need a headset (headphones + a microphone).

*FAQs = Frequently Asked Questions

6 **Match the pictures with the words.**

1 pair of scissors O
2 screwdriver .....
3 can of oil .....
4 scales .....
5 tape measure .....
6 thermometer .....
7 spanner .....
8 pair of goggles .....
9 funnel .....
10 key .....
11 magnifying glass .....
12 torch .....
13 switch .....
14 fork-lift truck .....
15 crate .....

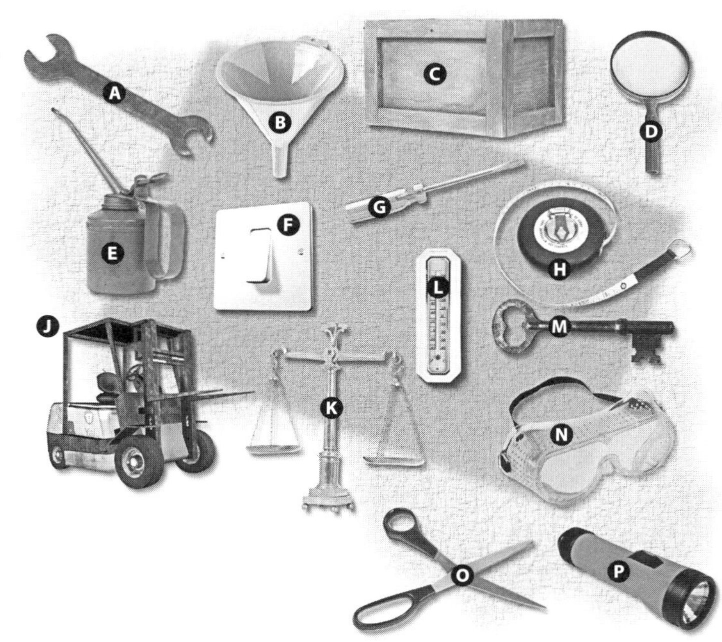

7 Write the name of the items you need to do these things.

	You want to ...	You need a ...		You want to ...	You need a ...
1	remove the screws on a cover	*screwdriver*	9	tighten some bolts	....
2	open a locked door	....	10	protect your eyes	....
3	look at something very small	....	11	weigh a parcel	....
4	move a very heavy box	....	12	lubricate a moving part	....
5	put some oil into a machine	....	13	turn off a machine	....
6	cut some paper	....	14	see in a dark place	....
7	pack an engine for delivery	....	15	measure a wall	....
8	check the temperature	....			

8 Read the first email. Complete the second email with the phrases in the list.

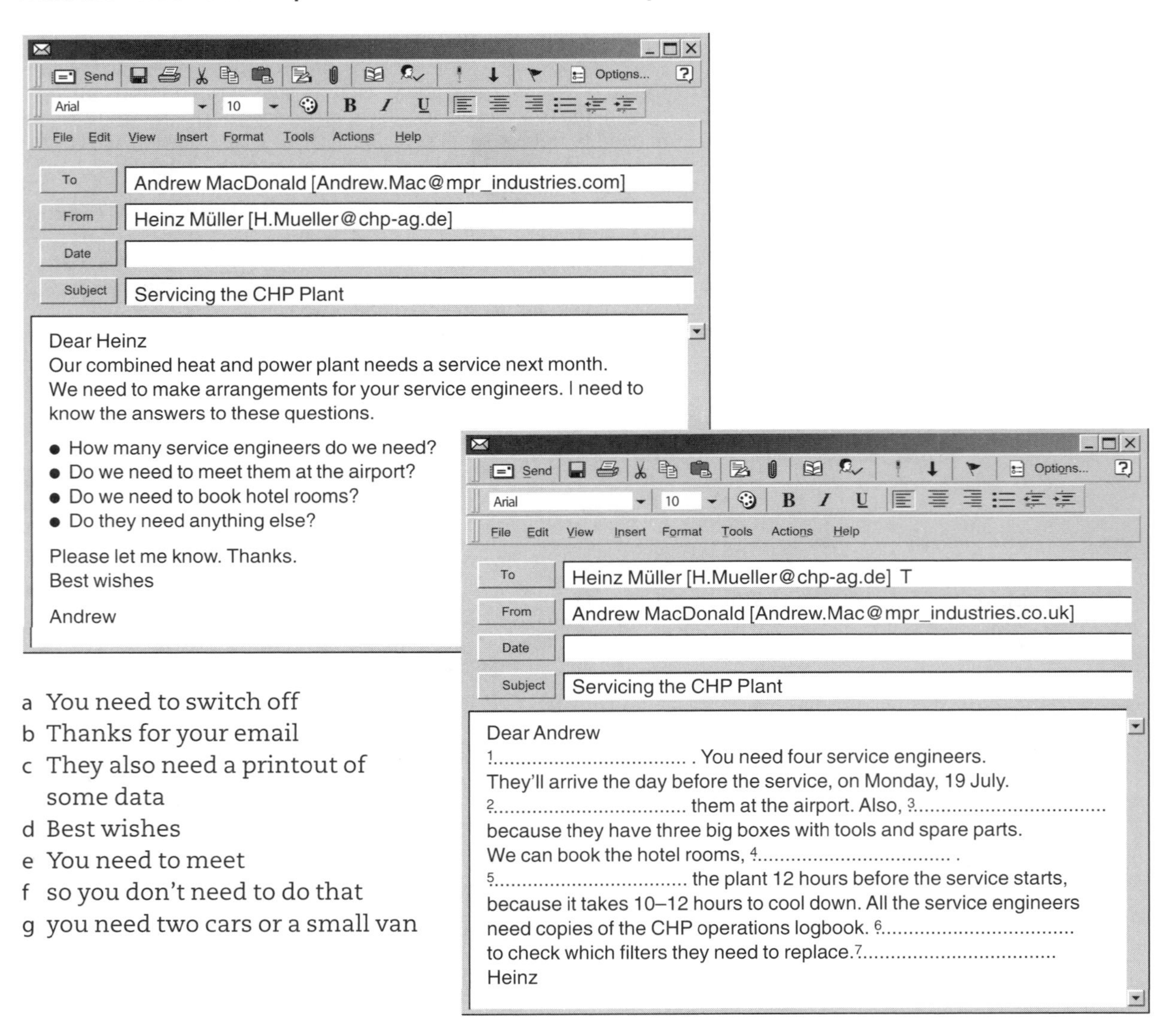

To: Andrew MacDonald [Andrew.Mac@mpr_industries.com]
From: Heinz Müller [H.Mueller@chp-ag.de]
Date:
Subject: Servicing the CHP Plant

Dear Heinz
Our combined heat and power plant needs a service next month.
We need to make arrangements for your service engineers. I need to know the answers to these questions.

- How many service engineers do we need?
- Do we need to meet them at the airport?
- Do we need to book hotel rooms?
- Do they need anything else?

Please let me know. Thanks.
Best wishes

Andrew

To: Heinz Müller [H.Mueller@chp-ag.de] T
From: Andrew MacDonald [Andrew.Mac@mpr_industries.co.uk]
Date:
Subject: Servicing the CHP Plant

Dear Andrew
1................................ . You need four service engineers.
They'll arrive the day before the service, on Monday, 19 July.
2................................ them at the airport. Also, 3................................
because they have three big boxes with tools and spare parts.
We can book the hotel rooms, 4................................ .
5................................ the plant 12 hours before the service starts, because it takes 10–12 hours to cool down. All the service engineers need copies of the CHP operations logbook. 6................................
to check which filters they need to replace. 7................................
Heinz

a You need to switch off
b Thanks for your email
c They also need a printout of some data
d Best wishes
e You need to meet
f so you don't need to do that
g you need two cars or a small van

Unit 9

1 Match the signs with the correct instructions.

1 Wear safety boots. *L*
2 Don't enter. .....
3 Don't use a mobile phone here. .....
4 Emergency exit this way. .....
5 Be careful when the floor is wet. .....
6 Don't touch. .....
7 Be careful! Dangerous liquid. .....
8 Wear safety goggles in this area. .....
9 Don't park here. .....
10 Be careful! Explosive material. .....
11 Watch out! Danger! .....
12 Don't switch on! .....
13 Danger of an electric shock. .....
14 Don't smoke here. .....
15 Don't wash. .....

2 Read the instructions for what to do in a thunderstorm. Write *Don't* before the instructions that are not true.

1 *Don't* stand under a big tree.
2 .... stand on top of a hill or high building.
3 .... get inside a car.
4 .... lie down on the ground.
5 .... use a telephone.
6 .... stay away from windows.
7 .... use electrical appliances (TVs, computers).
8 .... stay away from water (the sea, a lake).
9 .... carry metal objects.
10 .... panic!

3 What colour are these things? Write the colour under the picture.

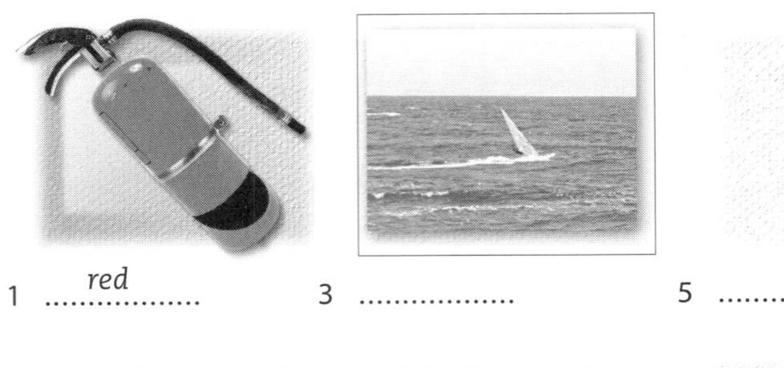

1 *red* .... 3 .... 5 .... 7 ....

2 .... 4 .... 6 .... 8 ....

4 Read the instructions and complete the sentences with *should* or *shouldn't*.

1 Switch off the machine first.
= The machine *should* be switched off first.

2 Don't connect the wire to terminal A.
= The wire *shouldn't* be connected to terminal A.

3 Clean the LCD display regularly.
= The LCD display .... be cleaned regularly.

4 Don't take food and drink into the control room.
= Food and drink .... be taken into the control room.

5 Don't remove the safety guard.
= The safety guard .... be removed.

6 Install the new sound card first.
= The new sound .... be installed first.

7 Change the password every day.
= The password .... be changed every day.

8 Don't store the cylinders outside.
= The cylinders .... be stored outside.

9 Replace the cover after use.
= The cover .... be replaced after use.

10 Don't connect the blue cable to the battery.
= The blue cable .... be connected to the battery.

5 Complete the questions with *Is*, *Are*, *Do*, or *Does*.

1 ...*Do*... you work for Fiat?
2 .... you a mechanical engineer?
3 .... he a customer?
4 .... she work here?
5 .... Sarah a computer technician or a programmer?
6 .... you know who Ali works for?
7 .... you a maintenance technician?
8 .... I need a special pass?
9 .... she know where to go?
10 .... he new here?

6 Read the questions and underline the right answer.

1 Do you work for Dunlop?	*No, I don't.* / No, I'm not.
2 Is Patricia a manager?	Yes, she does. / Yes, she is.
3 Do I know him?	Yes, you do. / Yes, you are.
4 Are you from RTC?	No, I don't. / No, I'm not.
5 Does Bill work for GE?	Yes, he does. / Yes, he is.
6 Is Maria Perrez from Brazil?	No, she doesn't. / No, she isn't.
7 Do you know what this does?	Yes, I do. / Yes, I am.
8 Does Ms Abdellah know Erika?	No, she doesn't. / No, she isn't.
9 Are you an electrician?	Yes, I do. / Yes, I am.
10 Does your company make electric motors?	Yes, it does. / Yes, it is.

7 Put the words in the right order to make questions.

1 a / you / mechanical engineer / are
Are *you a mechanical engineer*...?

2 an electronics company / work for / you / do
Do ....?

3 aerospace industry / in the / is / your company
Is ....?

4 phones / produce / does / your company / mobile
Does ....?

5 to have / a visitor's pass / you / do / need
Do ....?

6 the first floor / is / there / on / a photocopier
Is ....?

7 you / Nordea / Mr Linberg / are / from / in Oslo
Are ....?

8 Write answers to these questions.

1 Who do you work for? 2 What's your job? 3 Where do you work?

Unit 10

1 Match the items in the picture with the words in the list.

1 window *A*
2 keys .....
3 tool box .....
4 desk .....
5 radiator .....
6 tape measure .....
7 glasses .....
8 drawer .....
9 shelf .....
10 mobile phone .....
11 files .....
12 clock .....

2 Look at the picture again and complete the sentences with the words in the list.

behind between in *in front of* next to on under

1 The chair is *in front of* the desk.
2 The briefcase is .... the floor.
3 The pencils are .... the calculator and the PC.
4 The radiator is .... the desk.
5 The lamp is .... the phone.
6 The tape measure is .... the shelf, ... the files.
7 The newspaper is .... the briefcase.
8 The bottle is .... the mobile phone and the folder.
9 The photocopier is .... the shelves.
10 The keys are .... the chair.
11 The toolbox is .... the book shelves.
12 The telephone is .... the PC.

3 Angelo needs a meeting room. He phones Sheila. Look at the plan and underline the correct answer.

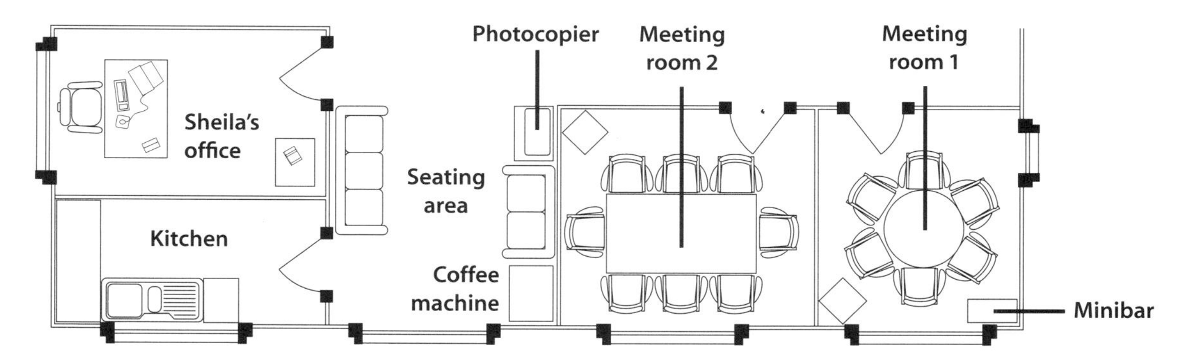

ANGELO Hello, Sheila. This is Angelo.
SHEILA Hi, Angelo. What's up?
ANGELO I need a meeting room with a minibar, a computer, and a photocopier.
SHEILA Is it for today?
ANGELO [1] <u>*Yes, it is.*</u> / *No, it isn't.* At four o'clock this afternoon.
SHEILA For how many people?
ANGELO Seven.
SHEILA OK. You can have room 2.
ANGELO Is the photocopier in room 2?
SHEILA [2] *Yes, it is.* / *No, it isn't.* But it's next to room 2.
ANGELO OK. And is the minibar in room 2?
SHEILA [3] *Yes, it is.* / *No, it isn't.* It's in room 1, but you can get coffee from the coffee machine.
ANGELO Is the coffee machine in the kitchen?
SHEILA [4] *Yes, it is.* / *No, it isn't.* It's in the seating area next to the kitchen.
ANGELO Is there a computer in room 2?
SHEILA [5] *Yes, there is.* / *No, there isn't.* But you can use my computer.
ANGELO All right. Thanks! Is the English dictionary in your office?
SHEILA [6] *Yes, it is.* / *No, it isn't.* It's on the small table on the left.

4 Match the times with the clocks.

A

B

C

D

E

F

G

H

1 It's five to one. *G*
2 It's eight fifteen. .....
3 It's half past four. .....
4 It's six twenty-five. .....
5 It's half past three. .....
6 It's three o'clock. .....
7 It's eleven forty-five. .....
8 It's ten to two. .....

5 Write the times a different way.

1 It's five to one. *It's twelve fifty-five.*
2 It's eight fifteen. ....
3 It's half past four. ....
4 It's six twenty-five. ....
5 It's half past three. ....
6 It's three o'clock. ....
7 It's eleven forty-five. ....
8 It's ten to two. ....

6 Read the information and write the times.

1 It's 4.30 in Berlin. New York is 5 hours behind Berlin. What's the time in New York? ....*11.30*....

2 You get up at 6.45. Breakfast takes half an hour. It takes you 20 minutes to drive to work. When do you arrive at work? ....

3 Your train leaves at 9.25. The journey takes 45 minutes. When do you arrive? ....

4 It is now 2.25 and you have a meeting. It takes 5 minutes to walk to the meeting. The meeting lasts an hour. Then you walk back to your office. What's the time when you arrive back in your office? ....

5 Your flight is at 2.30. You want to arrive at the airport one hour before the flight leaves. It takes 45 minutes to drive to the airport. When do you have to leave home? ....

6 You finish work at 5.30. It takes you 40 minutes to drive home. Today you want to stop on the way home and go shopping for an hour. When do you arrive home? ....

7 Complete the conversations with *this*, *that*, *these*, or *those*.

Unit 11

1 Complete the conversation with the words and phrases in the list.

can't (x 3) don't (x2) have wrong (x2) need to not

A I [1]..*can't*.. replace this engine control unit!
B Why [2]..................?
A There's a problem with the new control unit.
B What's [3].................. with it?
A The holes are in the [4].................. place.
B Why [5].................. you drill some new holes?
A What? I [6].................. do that!
B OK. Why [7].................. you repair the old control unit?
A That's a problem. I [8].................. replace the air-fuel chip.
B Do we [9].................. a new chip?
A Yes, but ...
B OK, so why don't you just replace the chip?
A We [10].................. do that! The customer wants a new unit.
B Of course we can!

2 Match the words with the pictures.

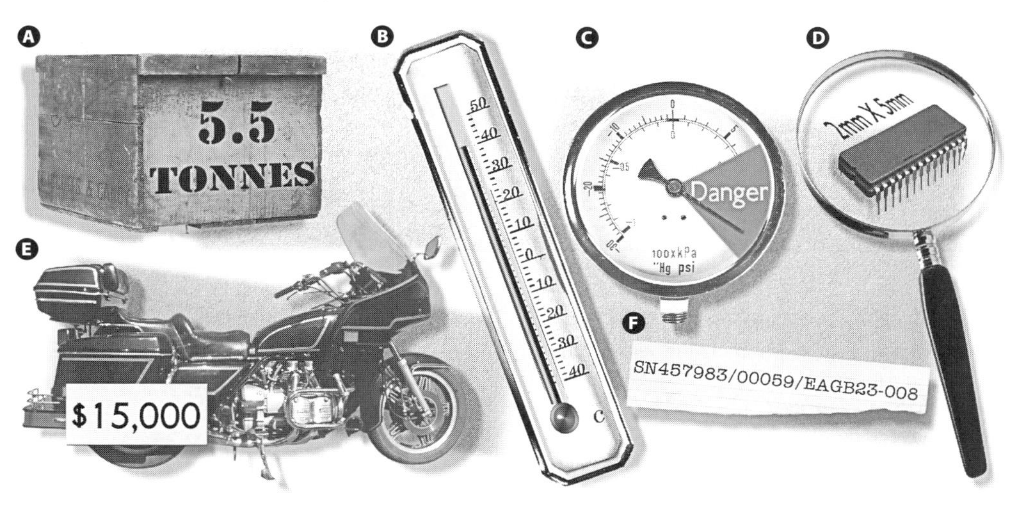

1 dangerous ..C..
2 expensive .....
3 hot .....
4 heavy .....
5 long .....
6 small .....

3 Complete the sentences with a phrase from the list.

it's too old	they're too small	*it's too expensive*
they're too difficult	they're too heavy	it's too far
it's too small	it's too dangerous	

1 We can't buy that laser printer – *it's too expensive* .
2 I can't walk to work from my home – ....
3 We can't use those screws – ....
4 I'm afraid we can't repair this computer – ....
5 I'm afraid we can't go into the construction area – ....
6 I can't lift these boxes – ....
7 I can't read this serial number – ....
8 I can't fill in these forms – ....

4 Match the instructions with the correct sign.

1 You have to stop. *E*
2 You can't park here. .....
3 You have to turn right. .....
4 You can't turn left. .....
5 You have to keep right or left. .....
6 You can't walk here. .....
7 You can't drive down this road. .....
8 You can't turn left or right. .....

5 Read this email about a visit to a company. Write four things Michael Kelly wants to do.

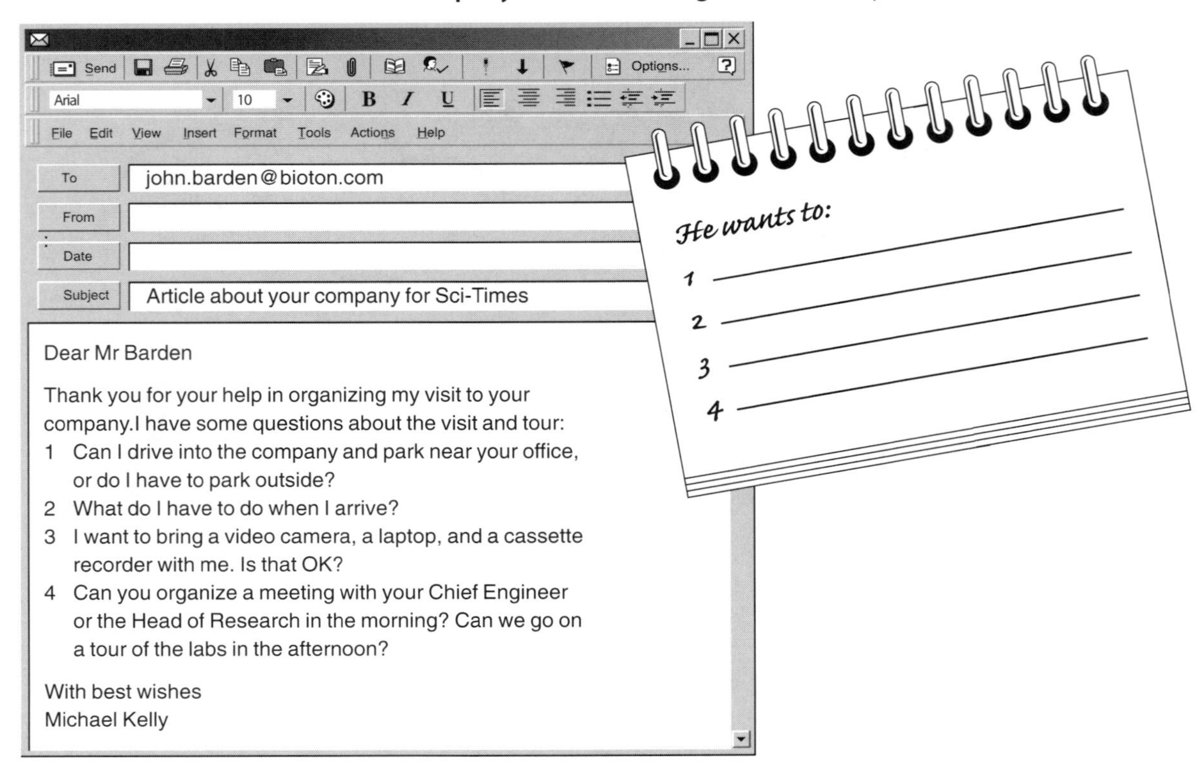

To: john.barden@bioton.com
From:
Date:
Subject: Article about your company for Sci-Times

Dear Mr Barden

Thank you for your help in organizing my visit to your company.I have some questions about the visit and tour:

1 Can I drive into the company and park near your office, or do I have to park outside?
2 What do I have to do when I arrive?
3 I want to bring a video camera, a laptop, and a cassette recorder with me. Is that OK?
4 Can you organize a meeting with your Chief Engineer or the Head of Research in the morning? Can we go on a tour of the labs in the afternoon?

With best wishes
Michael Kelly

6 Complete John Barden's reply with *can't* or *have to.*

To: michael.kelly@sci-times.co.uk
From:
Date:
Subject: Your visit

Dear Michael

Here are the answers to the questions you sent me about your visit. I'm afraid you 1...*can't*.......... drive into the company. You 2................ park in the visitors' car park.

You 3................ get a visitor's pass from reception. You 4................ enter any of the buildings without this pass.

I'm sorry, but you 5................ bring a video camera into the company. I think it's OK to bring a laptop, but I 6................ check with security.

I 7................ go to a meeting at one o'clock, so we 8................ go on a tour of the labs in the afternoon. We can do the tour in the morning and you can interview our Chief Engineer, Mr Cobain, in the afternoon.
I look forward to meeting you.
Kind regards
John Barden

7 Match the problems with the best solution.

1 *I don't have enough money.*
2 This PC is too slow for this program.
3 It's very hot in here.
4 I don't understand these instructions.
5 Is this cable long enough?
6 We don't have enough parts.
7 I don't know how to replace the sound card.

a Why don't you use a dictionary?
b Why don't you order some more?
c *Why don't you use a credit card?*
d Why don't you open the window?
e Why don't you measure it?
f Why don't you read the instructions?
g Why don't you upgrade it?

8 Read the instructions from The Plaza Hotel. Mark the statements true (T) or false (F).

1 You have to fill out a drinks card when you use the minibar. (T) / F
2 You can't have breakfast in the hotel after eleven o'clock in the morning. T / F
3 You can't have breakfast in your room. T / F
4 You have to reserve a table for dinner. T / F
5 You have to check out before twelve in the morning. T / F
6 You have to pay $5 a day to park your car. T / F
7 You can't load or unload your bags in front of the hotel. T / F
8 You can't smoke in your room. T / F

THE PLAZA HOTEL

Dear Guest
Welcome to The Plaza Hotel.

- If you use the minibar, please fill out the drinks card and take it to reception when you check out.
- Breakfast is served from 6.00 –10.30 a.m. in the Lincoln Room. If you would like breakfast served in your room, please fill out the breakfast card and leave it with reception.
- Dinner is served from 6.30–11.30 p.m. in the Roosevelt Grill on the second floor. Reservations are not necessary.
- Guests are asked to check out before 12.00 midday on the last day of their stay.
- The hotel car park is behind the hotel. Parking costs $5 a day. Guests can park their cars in front of the hotel for a short time to load or unload their bags.
- This is a non-smoking room. Please do not smoke in this room.
- Please do not leave money or credit cards in your room. Please leave them at reception.

We wish you a pleasant stay.

Unit 12

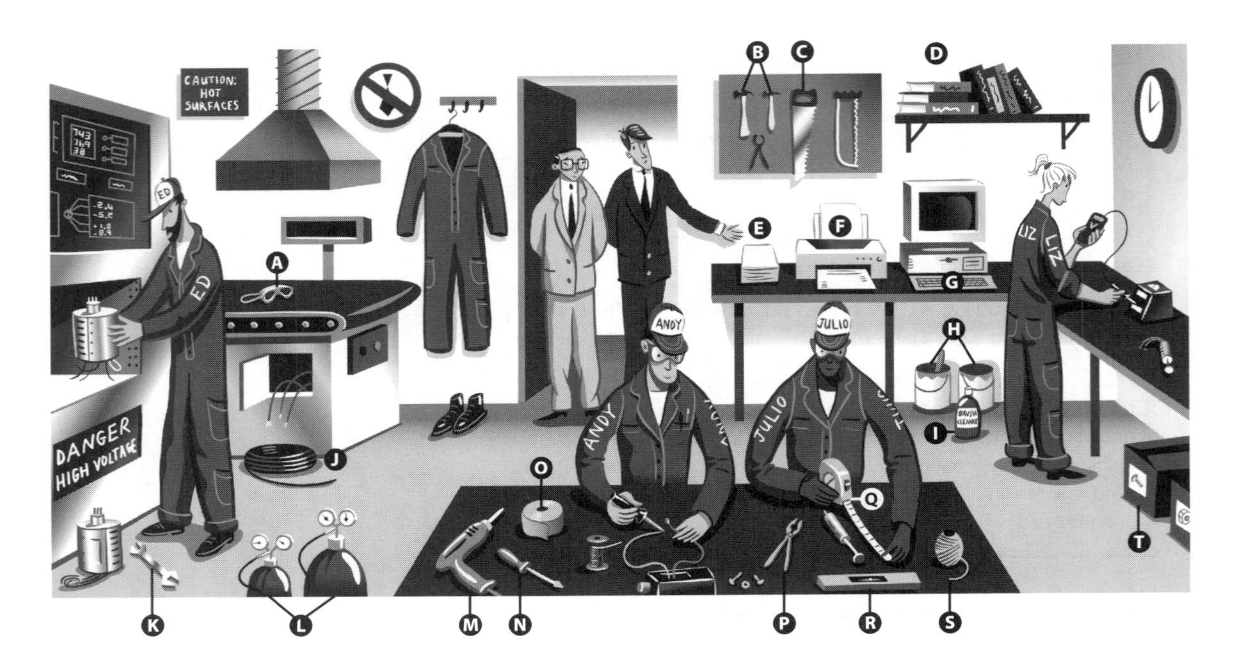

1 Match the words with the items in the picture.

1	tape measure	Q	6	drill	.....	11	tape	.....	16	cable	.....
2	pliers	.....	7	manuals	.....	12	paper	.....	17	gas	.....
3	spanner	.....	8	hammers	.....	13	paint	.....	18	string	.....
4	goggles	.....	9	spirit level	.....	14	boxes	.....	19	printer	.....
5	screwdriver	.....	10	saw	.....	15	bottle	.....	20	keyboard	.....

2 Erik Larsson is a production supervisor. Mr Funakoshi is a quality manager from Japan. He is visiting Erik's workshop. Complete Parts 1 and 2 of the conversation with the phrases in the list.

Part 1

's removing *are servicing* aren't using Is he repairing 's installing are they doing

ERIK LARSSON This is the workshop, Mr Funakoshi. We [1] *are servicing* the machines.

MR FUNAKOSHI I see. So you [2].................................. these machines today.

ERIK LARSSON That's right.

MR FUNAKOSHI I see. So, what [3].................................. exactly?

ERIK LARSSON Ed [4].................................. some parts from this machine.

MR FUNAKOSHI [5].................................. the parts?

ERIK LARSSON No, he [6].................................. new parts.

Part 2

're soldering are they wearing are you doing are repairing 'm testing

MR FUNAKOSHI	Excuse me. What [7]..................................?
LIZ	I [8].................................. some electrical parts with a voltmeter.
ERIK LARSSON	Over here, Andy and Julio [9].................................. some broken connections. They [10].................................. the wires to these terminals.
MR FUNAKOSHI	Why [11].................................. goggles?
ERIK LARSSON	Safety rules!

3 Use the words to make complete sentences.

1 I / install / some new software.
I'm installing some new software.

2 I / connect / the printer to the computer.

....

3 Be careful! You / stand / on the cables!

....

4 She / read / the manual / in English.

....

5 We / build / a new workshop.

....

6 They / test / the machine now. It / work / well.

....

4 Change the sentences into the negative form.

1 He's installing a new machine.
He's not installing a new machine.

2 Lorenzo's operating the machine today.

....

3 We're replacing the batteries.

....

4 She's doing the tests for Mr Nikolaev.

....

5 They're using the electric drill. It's working.

....

6 I'm checking the connections at the moment.

....

5 Put the words in the right order to make questions.

1 you / doing / are / what

What are you doing? ....?

2 she / it / adjusting / is

....?

3 are / what / soldering / you

....?

4 the / computer / they / are / changing / why

....?

5 the / leaking / are / batteries

....?

6 removing / he / the / cover / why / is

....?

6 Look at the items in 1 again. Which are countable and which are uncountable? Write the words in the boxes.

Countable	Uncountable
cables	*gas*

7 Complete the dialogues using *a*, *an*, *some*, or *any*.

1 A Is there ......*a*...... tape measure in the toolbox?

B Yes, but there isn't .... paper or .... pencil to write the measurements.

2 A Can I help you?

B Yes, there's .... white paint, ... paintbrush, and .... brush cleaner over there. Can you bring them over here?

3 A Is there .... gas in the bottle?

B No, it's empty. But there are .... full bottles in the storeroom.

4 A There are .... cables and .... printer in my office. Can you take them to the workshop, please?

B Sure.

5 A Are there .... tools in the car?

B Yes, there are .... wrenches, ... electric drill, and .... screwdriver.

A Is there .... petrol in it?

B Yes, it's full.

8 Complete the conversation with the phrases in the list.

a The key. Do you have it?
b But I don't have any money for a new tank.
c *I'm working on my motorbike.*
d I know. That's why I'm repairing it.
e There isn't any gasoline in it.
f Why?

A What are you doing?
B 1 *I'm working on my motorbike.*
A Why are you filling the fuel tank?
B 2 ....
A Look! The tank is leaking!
B 3 ....
A You can't repair it, you have to replace it!
B 4 ....
A What are you looking for?
B 5 ....
A No. I'm leaving!
B 6 ....
A It's too dangerous here.

9 In each group, one of the words is a thing *and* an action. Underline the actions.

1 *coffee*	*staple*	*calculator*	*manual*
2 bolt	tool	shelf	photo
3 woman	clothes	nail	scissors
4 cement	money	tank	chewing gum
5 oxygen	glue	ruler	credit card
6 information	problem	parking	screw

10 Complete the sentences with an action word from 9.

1 You can *staple* paper, but not bricks.
2 You can .... bricks, but not leather.
3 You can .... pieces of cardboard together, but not pieces of wool.
4 You can .... pieces of plastic together, but not pieces of polystyrene.
5 You can .... pieces of wood together, but not pieces of glass.
6 You can .... two pieces of metal together, but not two pieces of cotton.

Unit 13

1 Match the items on the left with a phrase on the right to make a sentence.

1 *A compass* — c
2 A fan
3 A lighter
4 Glue
5 Seat-belts
6 Spanners
7 A saw
8 Passwords
9 Goggles
10 A jack

a is for sticking things together.
b are for keeping you safe in a car.
c *is for finding the way.*
d are for tightening and loosening bolts.
e is for lifting a car.
f is for cooling an engine.
g are for protecting your computer.
h are for protecting your eyes.
i is for starting fires.
j is for cutting wood and metal.

2 Here are eight things you should wear or take with you when you go trekking in the desert. Rearrange the letters to make the words.

1 samcops c*ompass*
2 nussagles s.................
3 amp m.................
4 finke k.................
5 shilwet w.................
6 rorrim m.................
7 ath h.................
8 rightle l.................

3 Write five questions and answers about the things in 2.

Q *What's the compass for?*
A *It's for finding the way.*

1 Q ....?
A ....
2 Q ....?
A ....
3 Q ....?
A ....
4 Q ....?
A ....
5 Q ....?
A ....

4 Number the sentences in the conversations in the right order.

Conversation 1

a ..1.. ASSISTANT Can I help you, sir?

b ..... ASSISTANT You need some Supa-Glue. It's for sticking metal or plastic to glass.

c ..... ASSISTANT What do you want it for?

d ..... CUSTOMER I need to stick the mirror back on my car windscreen.

e ..... CUSTOMER Yes, I need some glue.

Conversation 2

a ..... KAI OK. Thanks for your help.

b ..... KAI How do I do that?

c ..... KAI I can't send this attachment to Yoshi. It's too big.

d ..1.. BEN What's the problem?

e ..... BEN Use the program Zip-Zipper. It's for making large files smaller.

f ..... BEN You need to compress the file.

Conversation 3

a ..... SVEN A remote sensor? What's it for?

b ..... SVEN OK, and where can I see what the temperature is?

c ..1.. SVEN What's that?

d ..... PETER In the control centre. The operators have to check that it doesn't get too hot, or too cold in here.

e ..... PETER It's a remote sensor.

f ..... PETER It's for measuring the temperature of this room.

5 Read the conversation and label the camera.

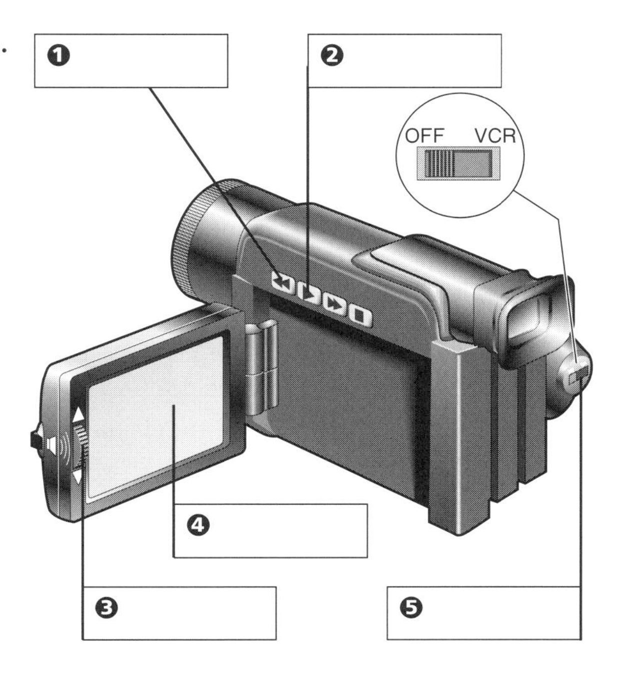

A Can we watch this video cassette on the camera?

B Yes, of course. First, you have to put the cassette into the camera.

A Right, that's easy.

B Now, open the LCD screen on the other side of the camera.

A OK. Now what?

B Turn the power switch from OFF to VCR. It's at the back, on the right.

A Where? Is it this switch?

B Yes, that's it.

A OK. The LCD screen's on. Now what?

B Press the rewind button.

A Is it this button at the top here?

B Yes, that's it. Wait until it rewinds to the beginning of the tape.

A ... OK! What do I do now?

B Press the PLAY button. It's next to the rewind button.

A Hey! Great! I can see the picture, but there's no sound.

B Turn up the volume then. The knob's on the side of the LCD screen.

6 Match the buttons to the functions.

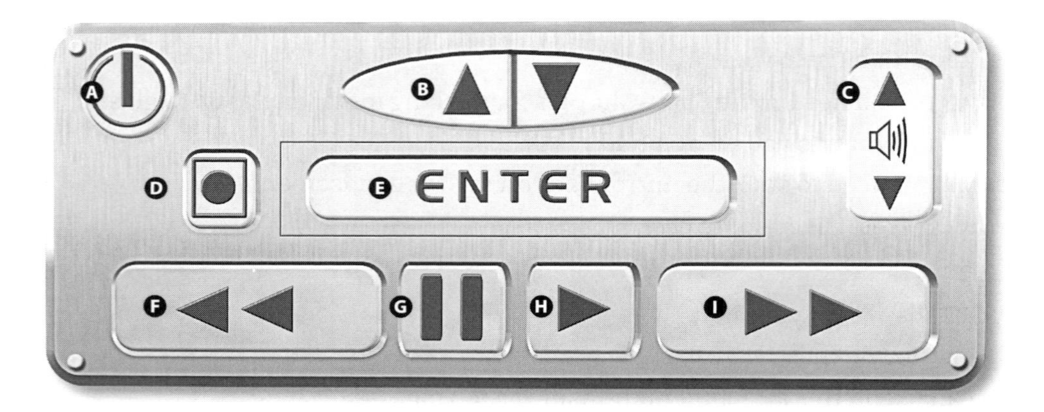

1 power switch ..A..
2 fast forward .....
3 pause .....
4 rewind .....
5 record .....
6 play .....
7 volume .....
8 move up / down the menu .....
9 select .....

7 Write the opposites of the verbs.

1 turn on *turn off / switch off*
2 increase ....
3 plug in ....
4 push ....
5 open ....
6 fast forward ....
7 pick up ....
8 stop ....

8 Complete the dialogue with the words in the list. You can use the words marked (x2) twice.

enter calls for how (x2) use button (x2) easy what

A [1] *How* do you switch the mobile on?
B Press the red [2].................. at the top.
A What's this button [3]..................?
B It selects the menu.
A How do you [4].................. new names and numbers?
B [5].................. the number keys.
A [6].................. does this knob do?
B It [7].................. the number that is displayed.
A And [8].................. do you switch the phone off?
B How do you think?
A Press the red [9].................. at the top!
B That's right!
A But that's [10]..................!

Unit 14

1 Match the items in A to the problems in B.

A

1 *screen*
2 diskette
3 boxes
4 two keys
5 CD
6 speaker
7 computer unit
8 manual

B

a broken
b missing
c dented
d *cracked*
e torn and dirty
f bent
g scratched
h crushed

2 These letters arrived in the quality department of a computer company. Complete the sentences with *was* or *were*.

❶ My computer arrived today, but it ...*was*... in very bad condition.

❷ I think the boxes .... crushed.

❸ *The side of the unit .... scratched and dented.*

❹ One of the speakers .... broken, and the software CDs .... scratched.

❺ The keyboard .... chipped, and two keys ....missing.

❻ *The screen .... cracked when it was delivered.*

❼ The other items .... in good condition, but the diskette for the printer .... bent.

❽ Pages 23 to 28 in the user manual .... torn, and it was difficult to read the instructions.

3 Write short conversations. Follow the example.

I / replaced / screen. (cracked)

A *I replaced the screen.*

B *What was wrong with it?*

A *It was cracked.*

We / changed / bolts. (bent)

A *We changed the bolts.*

B *What was wrong with them?*

A *They were bent.*

1 We / repaired / power cord. (worn)

....

.... ?

....

2 She / replaced / batteries. (leaking)

....

.... ?

....

3 I / changed / fuses. (burnt out)

....

.... ?

....

4 I / changed / tyre on the truck. (flat)

....

.... ?

....

5 We / painted / pipes. (rusty)

....

.... ?

....

6 I / fixed / handle on your bag. (broken)

....

.... ?

....

4 Read the report and mark the sentences true (T) or false (F).

FQ1 H - 05

Accident and damage report

Place: *Reception area (Helsinki workshop)*

Date of accident: *23rd December 2003*

Time: *09.45*

Description: *Fire*

Cause of accident: *The technician installing the computer network used a flammable cleaning liquid next to a gas heater. The windows were closed. A fire started.*

Damage: *The fire damaged the reception area. It melted the plastic seats, telephones, and some computer equipment. The high temperature cracked the mirrors and lights. It also broke the glass in the doors. The smoke damaged the materials on the walls and floor. The fire burnt the Christmas tree.*

Date: *12/23/03*

Name *HASSEL, Björn*

Signature: *Björn Hassel*

1 The accident happened on Christmas Day. T / (F)

2 The fire happened because of electrical problems. T / F

3 Björn Hassel signed the report on the day of the accident. T / F

4 It was dangerous to use the cleaning liquid because the windows were closed. T / F

5 The wallpaper is in good condition. T / F

6 The seats were not damaged. T / F

7 The heat cracked the mirrors. T / F

8 The Christmas tree melted. T / F

5 **Björn Hassel phoned his boss the next day to tell him about the accident. Complete the conversation with the words in the list.**

arrived *called* started contacted was (x 2) were used happened

BJÖRN	Hello, can I speak to Mr Hansen, please?
MR HANSEN	Speaking.
BJÖRN	This is Björn Hassel, in Helsinki. It's about the fire. I [1] *called* you yesterday, but you [2]................. out.
MR HANSEN	Yes, I was in Sweden. What [3].................?
BJÖRN	There was a fire in the Helsinki workshop. The technician [4]................. a flammable cleaning liquid on the computer. The windows were closed. The fire [5]................. because he used the liquid next to the gas heater.
MR HANSEN	What [6]................. damaged?
BJÖRN	Everything in the reception area.
MR HANSEN	I want to see it. I'll come tomorrow.
BJÖRN	I'm sorry, the office is closed tomorrow and Friday. Then it's the weekend. I [7]................. the insurance company this morning at nine o'clock. They [8]................. very quickly.
MR HANSEN	Good. But I want to speak to that technician!
BJÖRN	You don't need to speak to him, Mr Hansen.
MR HANSEN	Why not?
BJÖRN	It [9]................. me!

6 **Complete the text with *in*, *on*, or *at*.**

The fire happened [1] *in* December, [2]..... Christmas. The work started [3]..... the beginning of the week, [4]..... Monday. The accident happened [5]..... Tuesday the 23rd [6]..... nine forty-five. The window was closed because it's cold in Helsinki [7]..... winter. Mr Hansen was not there [8]..... Tuesday. Björn explained the accident to Mr Hansen [9]..... Wednesday the 24th. He contacted the insurance company [10]..... the morning. Mr Hansen didn't come to see the damage because the office was closed [11]..... the 25th and the 26th, and [12]..... the weekend.

7 **Write the Past Simple form of the words in the correct column.**

	1 syllable (●)	2 syllables (●●)		1 syllable (●)	2 syllables (●●)
1 move	*moved*		6 start		
2 paint		*painted*	7 sign		
3 test			8 ask		
4 use			9 bolt		
5 count			10 fix		

8 Write the dates in two ways.

	British English	American English
1 01/07	*the first of July*	*January seventh*
2 08/12	....	....
3 06/03	....	....
4 05/09	....	....
5 11/02	....	....
6 10/04	....	....

9 Read the text. Complete each part of the text with the words in the list.

The Leaning Tower of Pisa

began stopped *designed* increased finished started

The Tower of Pisa was [1] *designed* by the Italian architect, Bonanno Pisano. Construction work [2].................. in 1173. A few years later, the tower [3].................. to lean to one side. The building work [4].................. and started many times, and the tower was officially [5].................. in 1370. Over the next 600 years, the angle of the lean [6].................. .

added installed used decided

In the 1990s, some engineers [7].................. to correct the lean. They [8].................. large weights made of lead* (Pb) at the bottom of the tower to stop it moving. Then they [9].................. liquid nitrogen to freeze the foundations, but the tower started to lean again. They quickly [10].................. more lead, and the lean stopped.

installed closed collapsed* removed signed

In 1998, another old tower in Italy [11].................., and the Italian government [12].................. the Tower of Pisa. They [13].................. a contract with a British company to straighten the tower. The engineers [14].................. steel cables to hold the tower in place. Then they [15].................. earth from under the high side of the tower. The tower started to straighten.

completed visited corrected

The British engineers [16].................. the lean by 43 centimetres. On the 6th of June 2001, they [17].................. the project, and the government re-opened the tower. In November, tourists [18].................. the tower for the first time since 1999.

lead a very heavy metal (Pb) **collapse** to fall down

Unit 15

1 Complete the information in the text and the table.

Jim Shan is from Cincinnati in [1]................. He's a [2]................. .... for a company that makes petrochemicals. When Jim isn't working he likes watching baseball and playing golf. His favourite baseball team is the Cincinnati Reds. At the weekend Jim goes to [3]................. .... Golf Club to play golf. Jim speaks English and [4]................. .

Inga Telleffson is from [5]................. in Norway. She is a mechanical engineer at a small company that makes parts for ships' motors. Inga is married and has [6]................. children, Peder and Ragnhild. Inga speaks [7]................. languages: [8]................. , English, French, and [9]................. . She likes travelling when she has enough time. When she isn't working she likes [10]................. and [11]................. . In winter she especially likes skiing.

Name	Jim Shan	Inga Telleffson
Nationality	American	[16].................
Home town / city	Cincinnati, USA	Bergen
Profession	Lab Assistant	Mechanical Engineer
Languages	[12]................. , Chinese	Norwegian, [17]................. , [18]................. and German.
Hobbies	[13]................. and [14].................	Sailing, [19]................., walking – and [20]................. when she has enough time.
Other information	Favourite team: [15]................. Golf club: Elks Run	Two children, [21]................. and [22]................. .

2 Look at the questions. Tick (✔) the ones that are right and correct the ones that are wrong.

1 Where are you from? ✔...
2 ~~What do you make at the weekend?~~ *What do you do at the weekend?*
3 What is your first name? ....
4 How many languages speak you? ....
5 From where is Kai? ....
6 What are your hobbies? ....
7 Where are you live? ....
8 What sports you like? ....
9 What's your favourite sport? ....
10 How many children you have? ....

3 **Write the questions to these answers.**

1 *Where is Inga Telleffson from?* ? She's from Norway.
2 .... ? She lives in Bergen.
3 .... ? She's a mechanical engineer.
4 .... ? Four – Norwegian, English, French, and German.
5 .... ? Two.
6 .... ? Peder and Ragnhild.
7 .... ? Sailing, skiing, and walking.
8 .... ? No, she doesn't play golf.

4 **Write some similar questions to ask about Jim Shan. Ask about:**

- his home town
- his company
- his hobbies
- his job
- languages
- the weekends

5 **Match the fractions and percentages (1–10) with the words (a–j).**

1 50%	a five-eighths
2 33.3%	b three-fifths
3 75%	c five-sixths
4 25%	d two-thirds
5 60%	e one tenth
6 ⅔	f *a half*
7 ⅝	g a third
8 ⅞	h a quarter
9 ⅒	i three-quarters
10 ⅚	j seven-eighths

6 **Complete the text. Write the percentages as fractions, using the words in the list.**

three-quarters two-thirds half a quarter a fifth

Language skills not used by world's businesses

About [1] *half* (50%) of the people we asked can speak other languages. [2]__________(75%) of the people work in companies that use more than one language. Over [3]__________(50%) of their customers and colleagues are from a different country and speak a different language. Just under [4]__________(66%) work for companies that do business internationally.

Only [5]__________(25%) of the people say that language skills are not important in their workplace. About [6]__________(20%) say their companies lose business because of communication problems with people from different countries.

Over [7]__________(50%) are learning a language. After English, Spanish is the favourite language to learn.

7 **Read the text again. Mark the sentences true (T) or false (F).**

1 About half the people asked say they can speak a foreign language. (T) / F
2 Not many of the people asked work in multilingual workplaces. T / F
3 A lot of the people asked say their colleagues and customers are from other countries. T / F
4 Just over a third of the people asked say their companies do business internationally. T / F
5 About three-quarters of the people asked say language skills are important at work. T / F
6 About 50 of the 534 people asked say their companies lose business because of language problems. T / F
7 Half the people asked say they are learning English. T / F
8 Spanish is the top foreign language to learn. T / F

8 **Choose the best answer for each question, a, b, or c.**

1 **Rolls Royce Motor Cars Ltd. is owned by a** *German* **company.**

a *German*
b British
c American

2 **Over 60% of the Swiss speak German as a first language, about 20% speak French and just over 6.5% speak ....**

a Spanish
b Italian
c Romansh

3 **You can eat nasi lamak in a .... restaurant.**

a Malaysian
b French
c Greek

4 **Nokia is a .... company.**

a Japanese
b Finnish
c Danish

5 **................. of the Earth is covered by water.**

a 50%
b 70%
c 90%

6 **................. use 24,602 kWh of electricity per person a year, Americans use 12,407 kWh a year and the Germans use 5,963 kWh a year.**

a Canadians
b Belgians
c Norwegians

7 **................. imports more per person than any other country in the world.**

a Hong Kong
b The Netherlands
c Mexico

8 **The United Kingdom has about 20 million Internet users, but there are 148 million Internet users in ....**

a Brazil
b Japan
c the United States

9 **The top three countries for mobile phones are: the United States, about 70 million; China, about 65 million; and ...., about 64 million.**

a Japan
b Italy
c India

10 **Which of these countries has no oil industry?**

a Venezuela
b Ecuador
c Panama

11 **________% of Canadians live in towns or cities.**

a 75%
b 50%
c 25%

12 **________ of a person's body is water.**

a A quarter
b Half
c Two-thirds

Unit 16

1 Complete the questions with *Can I* or *Can you*.

1 ......*Can you*...... tell me your name?
2 .... help me to check this control unit?
3 .... use your fax?
4 .... give me a 10 mm wrench?
5 .... park my car in Mr Lee's place?
6 .... send me a copy of the manual?
7 .... pay with my credit card?
8 .... eat my sandwich here?

2 What can you say in these situations? Write questions using *Can I* or *Can you*.

1 You want someone to fill in a form.
Can you fill in this form, please ?
2 You want to know if it's OK to use the phone.
.... ?
3 The air-conditioning is making a lot of noise. You want someone to turn it off.
.... ?
4 You want to put your tools in someone's car.
.... ?
5 You need to insulate a wire. You want to know if it's OK to use tape.
.... ?
6 You want someone to check some calculations.
.... ?
7 You want to leave your bag at reception.
.... ?
8 You want a technician to finish a job today.
.... ?

3 Two of these phrases mean *No*. The others mean *Yes*. Find the phrases that mean *No*.

1 Sure.
2 Sorry, but ...
3 Of course.
4 No problem.
5 I'm afraid ...
6 Help yourself.
7 Go ahead.
8 I'll do it right away.

4 **Match the questions (1–8) with the answers (a–h).**

1 *Can I smoke in the office?* — f
2 Can you give me Rajan's email address?
3 Can I sit down?
4 Can you help me with this form?
5 Can I phone my office in Amsterdam?
6 Can I check my emails?
7 Can you lend me some change?
8 Can I have a glass of water?

a Yes, of course. Are you tired?
b Sure. How much do you need?
c Sorry, but the email server isn't working.
d Sure, go ahead. There's a phone over there.
e Yes, of course. There's a water fountain in the workshop.
f *No, but there's a smoking area in the restaurant.*
g Sorry, but I'm really busy. Ask Lisa.
h I'm afraid I can't remember it.

5 **Write short conversations. Use the questions and answers in 4 to help you.**

borrow / screwdriver?
yes
Can I borrow your screwdriver?
Sure, go ahead.
Thanks.

borrow / calculator?
no / no batteries
Can I borrow your calculator?
I'm afraid there are no batteries in it.
OK, it doesn't matter.

1 lend / mobile phone?
no / not charged
....
....
....

2 look at your manual ?
yes / on the shelf
....
....
....

3 borrow / English dictionary?
no / Stefan is using it
....
....
....

4 use / your car this afternoon?
no / no petrol in the tank
....
....
....

5 use / the photocopier?
yes / you need a code – 3039 + *Enter*
....
....
....

6 give / change for a coffee?
yes / here's two euros
....
....
....

6 **A manager in a sweet factory is phoning a production supervisor. Read the conversation and complete the specification sheet.**

A Tomas, I need to know how long it takes to do certain jobs. Can you help me?
B Yes, of course.
A Thanks. OK, how long does it take to fill 5,000 boxes of chocolates?
B With the FT600, it takes about ten minutes.
A OK. And how long does it take to wrap* 15,000 sweets with the GD 1100?
B Approximately fifteen minutes.
But we're servicing it. You can use the GD 850, but it's not as fast.
A How long does it take if you use the GD 850?
B The machine can do 750 per minute.
A OK, so that's about twenty minutes.
B That's right.
A One more question – how long does it take to wrap chocolate eggs?
B We can do 700 per minute. We use the GD 880.
A OK. Thanks for your help!
B You're welcome!

wrap = to put paper on sweets or other products

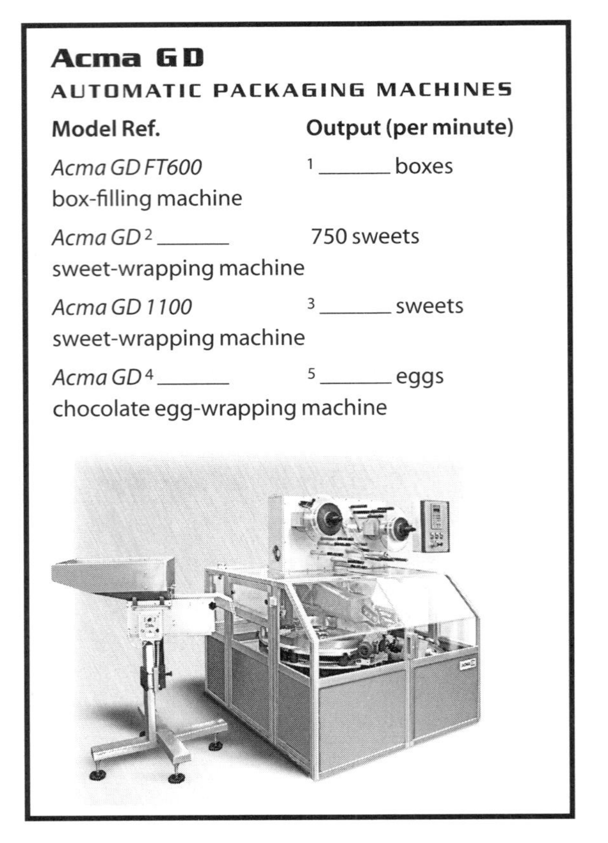

Acma GD
AUTOMATIC PACKAGING MACHINES

Model Ref.	Output (per minute)
Acma GD FT600 box-filling machine	1 ______ boxes
Acma GD 2 ______ sweet-wrapping machine	750 sweets
Acma GD 1100 sweet-wrapping machine	3 ______ sweets
Acma GD 4 ______ chocolate egg-wrapping machine	5 ______ eggs

7 **Look at the table and complete the conversations.**

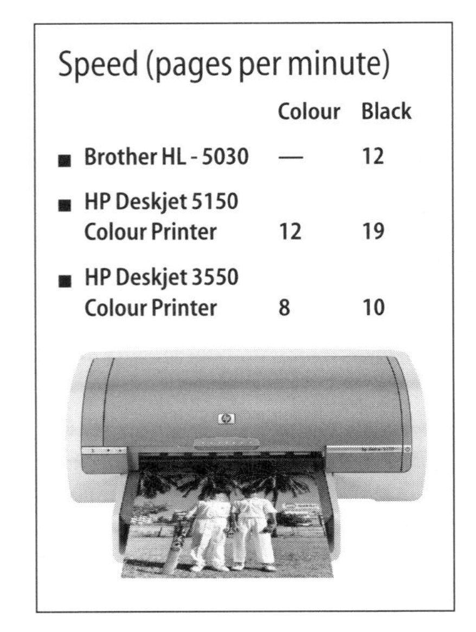

Speed (pages per minute)

	Colour	Black
■ Brother HL - 5030	—	12
■ HP Deskjet 5150 Colour Printer	12	19
■ HP Deskjet 3550 Colour Printer	8	10

A How long 1 *does* it 2................. to print 24 pages with the Brother HL-5030?
B It 3................. two minutes.
A OK, so 4................. .... does it take using an HP 5150?
B It prints twelve 5................. per minute too, but in colour!
A 6................. about with the HP 3550?
B How many pages do you want to print?
A 25 pages, in black.
B 7................. .... two and a half minutes.
A And if we use the HP 5150, 8................. .... to print 24 pages in colour?
B Two minutes.
A And what 9................. in black?
B It prints nineteen 10................. per minute. Do you want me to do the calculation for you?
A No, thanks. 11................. .... does it take to change the cartridge?
B I don't know!

8 Match the beginnings of the phrases with the endings.

1 *a tube of* — f
2 a roll of
3 a bottle of
4 a packet of
5 a piece of
6 a can of
7 a bag of
8 a barrel of
9 a box of
10 a sheet of
11 a glass of

a polystyrene / chocolate
b oil / beer
c Diet Cola / motor oil / soup
d diskettes / blank CDs / staples
e water / orange juice
f *glue / toothpaste / antiseptic cream*
g paper / cardboard / glass
h tape / film
i wine / alcohol / mineral water
j cement / sand
k screws / chewing gum / washers

9 Which words are countable, and which words are uncountable? Write C or U.

1 *U* money
2 ..... machines
3 ..... people
4 ..... equipment
5 ..... aspirins
6 ..... work
7 ..... oil
8 ..... batteries

10 Complete the sentences with *much* or *many*.

1 How *much* money do you need?
2 How .... work is there to do today?
3 How .... batteries does your phone need?
4 How .... equipment do we need for the job?
5 How .... aspirins are there in the first aid kit?
6 How .... oil is there in the tank?
7 How .... machines can you service in one day?
8 How .... people work in the company?

11 Read the calculations. Put a tick (✓) if a calculation is correct, and a cross (x) if it's wrong.

1 Eight divided by four equals four. ☒
2 Eighty-one minus eleven equals seventeen. ☐
3 Fourteen multiplied by two equals twenty plus eight. ☐
4 Thirty-three divided by eleven equals forty-seven minus forty-four. ☐
5 One hundred and twenty-five multiplied by two point five equals two hundred and fifty. ☐
6 Seven plus three equals ten, divided by three equals thirty. ☐

Unit 17

1 Match the pictures with the words.

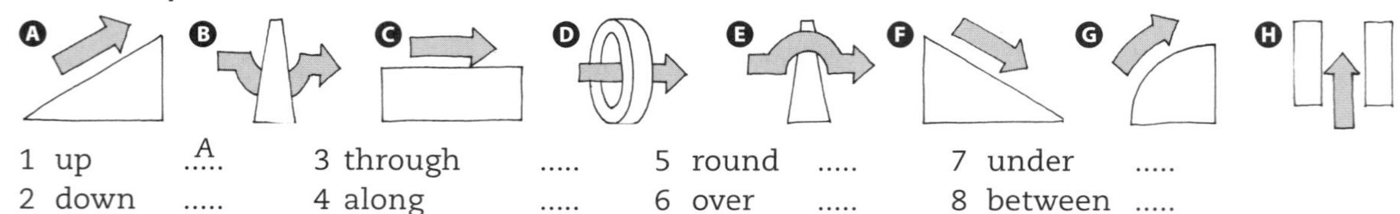

1 up	*A*	3 through	.....	5 round	.....	7 under	.....
2 down	.....	4 along	.....	6 over	.....	8 between	.....

2 Complete the sentences with the words in 1.

1 You can park the car over there. There is a space *between* those two cars.
2 From reception, go .... the stairs to the third floor and then turn left.
3 We like going for walks .... the beach.
4 We can't ski .... there. It's too dangerous.
5 You can't drive .... the bridge when there's a very strong wind.
6 There's a tunnel .... the river for cars and trains.
7 A security guard walks .... the factory every hour to check that everything is OK.
8 We can't get the machine .... the door. It's too big!

3 Read the instructions and draw:

- the pipeline
- the electricity cable
- the road
- the railway line.

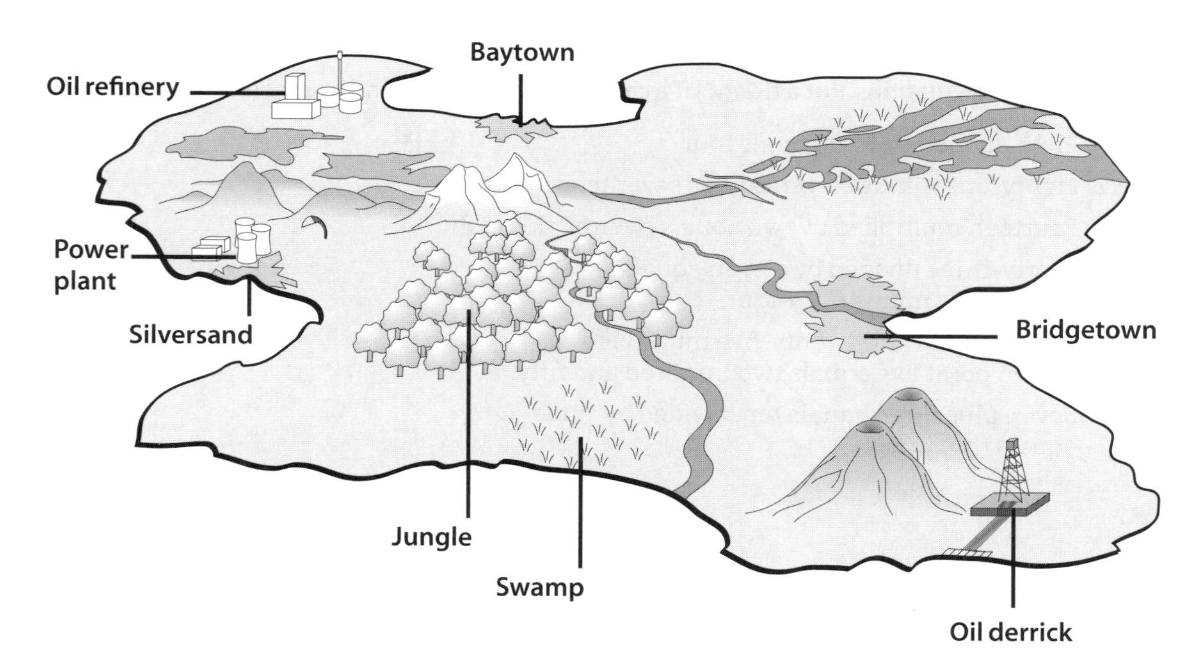

1 The pipeline starts at the oil derrick. It runs along the coast, over a river, through a swamp and through the jungle to Silversand. From Silversand it goes through a tunnel under a hill. Then it runs between the two lakes to the oil refinery.
2 The electricity cable starts at the power plant in Silversand. It runs over the pipeline, through the jungle, over the road and the river, and then to Bridgetown.
3 The road from Baytown runs round the hill on the right. Then it goes along the river, through the jungle. When the road comes out of the jungle, it goes over a bridge over the river. Then it goes under the railway line and between the two volcanoes to the oil derrick.
4 The railway line starts at Bridgetown. It runs over two bridges over the road and the river. It goes round the jungle, and between the jungle and the swamp. Then it goes along the coast to Silversand.

4 Complete the conversations with *come* / *coming* or *go* / *going*.

Conversation 1

A Where are you [1]......*going*......?

B I'm [2]................. out for a coffee. Do you want to [3]................. with me?

A No thanks. I have to [4]................. to a meeting.

Conversation 2

B Hello, David. This is Etienne Morin.

A Hi, Etienne. Where are you?

B I'm at the station. Can you [5]................. and get me?

A Yes, no problem. [6]................. to the bus stop in front of the station and wait for me there.

B OK. That's great. Thanks, David.

Conversation 3

A Excuse me, does the 53 bus [7]................. to the station?

B No, it doesn't. That's the airport bus. The 35 and 50 buses [8]................. to the station.

A OK. I'm [9]................. to the station now. I'll be back at two-thirty.

B OK, but when you [10]................. back, don't forget to call Tom.

5 Where would you say or hear these things? Tick (✔) the correct box.

	at a station	at an airport	in a taxi
1 *When does the train leave?*	☑	☐	☐
2 Can I have a receipt, please?	☐	☐	☐
3 Can I see your boarding card?	☐	☐	☐
4 This is the last call for Mr Kataki.	☐	☐	☐
5 Single or return?	☐	☐	☐
6 It leaves from platform one.	☐	☐	☐
7 Did you pack this bag yourself?	☐	☐	☐
8 Follow that tuk-tuk!	☐	☐	☐

6 Complete the conversations with the phrases in the lists.

Conversation 1

Have a good flight. How many bags do you have? Did you pack them yourself?

A Can I see your ticket and your passport, please?
B Yes, of course. Here you are.
A Thank you. [1].. .
B Two.
A [2].. .
B Yes.
A OK. Here's your boarding card. [3].. .
B Thank you.

Conversation 2

When does the train leave? At five to eight. A ticket to Chicago, please.

A [4].. .
B One way or a round trip, sir?
A A round trip.
B OK, that's forty-five dollars, please.
A [5].. .
B In twenty minutes. [6].. .
A And from which platform?
B Platform six.

Conversation 3

Do you need a hand with your bags? And I need a receipt. How much is that?

A OK, we're here. The Omni Chicago Hotel.
B Thanks. [7]... .
A That's $17.40, please.
B OK. [8]... . Make it for twenty dollars.
A Thank you, sir. Here's your receipt.
B Thanks.
A [9].. .
B No, thanks. I can manage.

7 Say these words to yourself. Circle the letters that are silent.

1 receipt
2 guarantee
3 knob
4 wrong
5 gauge
6 hour
7 wrench
8 scissors
9 adjust

8 **Look at the map and read the directions. Write the name of the building you arrive at.**

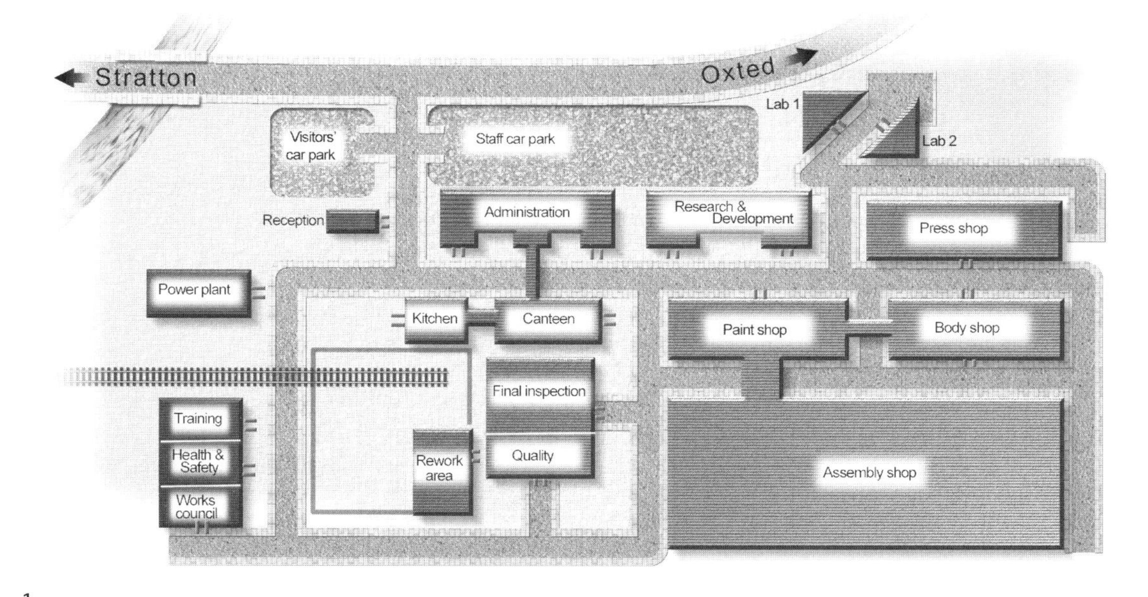

1 ....

The factory is 10 kilometres from Stratton. Go over the bridge and take the first road on the right. The visitors' car park is on the right-hand side. Park your car there and go to reception and get a visitor's pass.

After you have your visitor's pass, walk down the road past the administration building. At the end of the road, turn right and follow the road. It turns left at the power plant. Go over the railway line. My office is in the second building on the right.

2 ....

Please come to my office after your training course.

Go out of the training building and turn right (don't cross the railway line). At the end of the road, turn left. Walk past the rework area and quality building, then turn left. Take the first road on the left. My office is on the first floor of this building.

3 ....

Please come to my office after lunch.

When you leave the canteen, turn left and walk down the road, then turn right at the end of the road. Walk down the road between the paint shop and the R&D building. Turn left before the press shop and body shop. Follow the road and keep left. My office is on the third floor of the triangular building on the right-hand side.

4 ....

After your meeting with the works council, go out of their building and turn left. Take the first road on your left, walk up the road over the railway lines, and follow the road to the right. You walk past the road that goes to the reception and go under the passage between the administration building and canteen. My office is in the building after the administration building. There are two entrances to this building; take the second one. My office is on the ground floor.

5 ....

It's raining, so the best way there is to go to the second floor and use the passage to cross from here into the paint shop. Then we go down to the ground floor and use the exit opposite the canteen. We cross over to the canteen, go to the second floor again, and use the passage to cross over. Brian's office is on that floor.

Unit 18

1 Read the information. Match the instructions with the pictures.

How to clean your mouse

A computer mouse can be difficult to use after a few months of use.
Your mouse needs to be cleaned regularly.

a Remove the plastic ring.
b Replace the ring.
c Put the ball back into the mouse.
d Clean the small rollers with a cotton swab and some alcohol.
e Remove the ball from the mouse.
f Turn the mouse upside down.
g Wash the ball in warm water, and dry it with a soft cloth.
h Let the rollers dry completely.
i Lock the ball into place by turning the ring clockwise.
j Turn the plastic ring anti-clockwise.

❶ ❷ ❸ ❹ ❺

ALCOHOL

❻ ❼ ❽ ❾ ❿

2 Put the instructions for cleaning a mouse in the correct order.

1 ..*f*... 2 ..... 3 ..... 4 ..... 5 ..... 6 ..... 7 ..... 8 ..... 9 ..... 10 .....

3 Complete the instructions. Use the words in the lists.

measure loosen Remove Position Take off *Use* cut

How to change a British plug

Part 1

1 ...*Use*... a screwdriver to 2.................. the large screw on the new plug. 3.................. the cover.

4.................. the cable over the plug to 5.................. the length you need.

Use a knife to 6.................. the plastic insulation on the cable.

7.................. the plastic insulation, so that you can see the wires.

connect fix check Replace Locate switch on tighten

Part 2

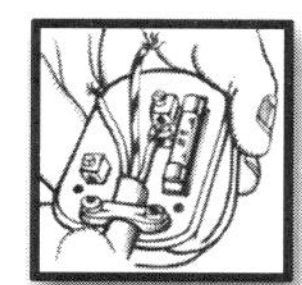

[8].................. the correct terminals and [9].................. the wires to the terminals as follows:
— the blue wire to the neutral terminal (A)
— the green and yellow wire to the earth terminal (B)
— the brown wire to the live terminal. (C)

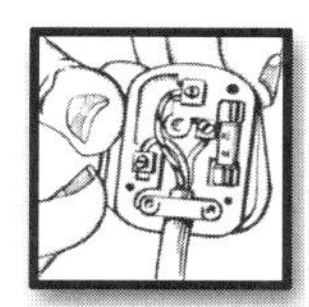

To [10].................. the wires in position, [11].................. the screws of the terminals. Next, [12].................. that the correct fuse is fitted to the plug.

[13].................. the cover and tighten all the screws. Finally, plug in the appliance and [14].................. it ....

4 Separate these two-part verbs using *it* or *them*.

1 Lift up the boxes. *Lift them up*
2 Turn on the lights. ....
3 Fill in the order form. ....
4 Hold down the springs. ....
5 Slow down the machine. ....
6 Take off the wheels. ....
7 Key in the code and the password. ....
8 Switch off the power. ....
9 Take off your face mask. ....
10 Put on your gloves. ....

5 Complete the sentences with the verb in brackets. Use *Do X to do Y* or *Do X by doing Y*.

1 Press the OFF button *to stop* the machine. (stop)
2 Raise the cover *by pulling* the lever. (pull)
3 Set the correct date .... button D. (press)
4 Press button T .... the time. (set)
5 Use a screwdriver .... the container. (open)
6 Disconnect the car battery .... the terminals. (remove)
7 Turn the knob clockwise .... the power. (switch on)
8 Unlock the door .... the key anti-clockwise. (turn)
9 Press and hold the switch down .... air pressure. (increase)
10 Activate the fire alarm .... the glass. (break)

6 Underline the best answer.

What happens if you:

1 drop a light bulb ? It *breaks* / *burns* / *bounces*.
2 put a burning match under a piece of paper? It *explodes* / *rusts* / *burns*.
3 smoke a cigarette in a room full of gas? It *melts* / *explodes* / *expands*.
4 put your ice-cream on a hot radiator? It *freezes* / *melts* / *bounces*.
5 leave petrol in the open air? It *expands* / *boils* / *evaporates*.
6 put too much air pressure in a car tyre? It *bursts* / *melts* / *breaks*.
7 put hot metal in cold water? It *dissolves* / *contracts* / *expands*.
8 leave your bicycle in the rain? It *freezes* / *rusts* / *stretches*.

7 Complete the description of how the Vacu Vin works. Use the words in the list.

place lowers opens removes pressure pump *put* press

HELMUT What's that?
SERGE It's a Vacu Vin wine preserver. It's for keeping wine after you open the bottle.
HELMUT How does it work?
SERGE Sometimes you start a bottle of wine, but you don't finish it. You want to keep the wine for another day.
HELMUT Right.
SERGE This is what you do. First, you [1] *put* the rubber stopper in the neck of the bottle, and [2]................. the pump over the stopper. Then you [3]................. the handle up and down three or four times. This [4]................. the air pressure inside the bottle. The pump [5]................. the air from the bottle through a small hole in the rubber stopper. The air can get out, but it can't get in. The [6]................. outside the bottle keeps the stopper closed. The wine doesn't oxygenate*, and it stays good for several days. You don't have to throw it away!
HELMUT That's a great idea! What happens if you want to open it again?
SERGE You just [7]................. the two sides of the stopper, and the hole [8].................. This lets the air back into the bottle. Then you can remove the stopper and drink the wine.

oxygenate = to react with oxygen in the air

8 Choose the best answer.

1 The Vacu Vin is a device for:
 a opening bottles of wine.
 b keeping wine which is already open.
 c removing wine from the bottle.

2 How does the Vacu Vin work?
 a The air is removed from the bottle so that the wine stays fresh.
 b The pump loosens the stopper in the neck of the bottle.
 c The air inside the bottle is replaced by air from outside.

3 You open the stopper:
 a with a tool.
 b with your fingers.
 c with the pump.

Unit 19

1 This bicycle needs repairing! Look at the picture and complete the note with words from the list. Use each word once.

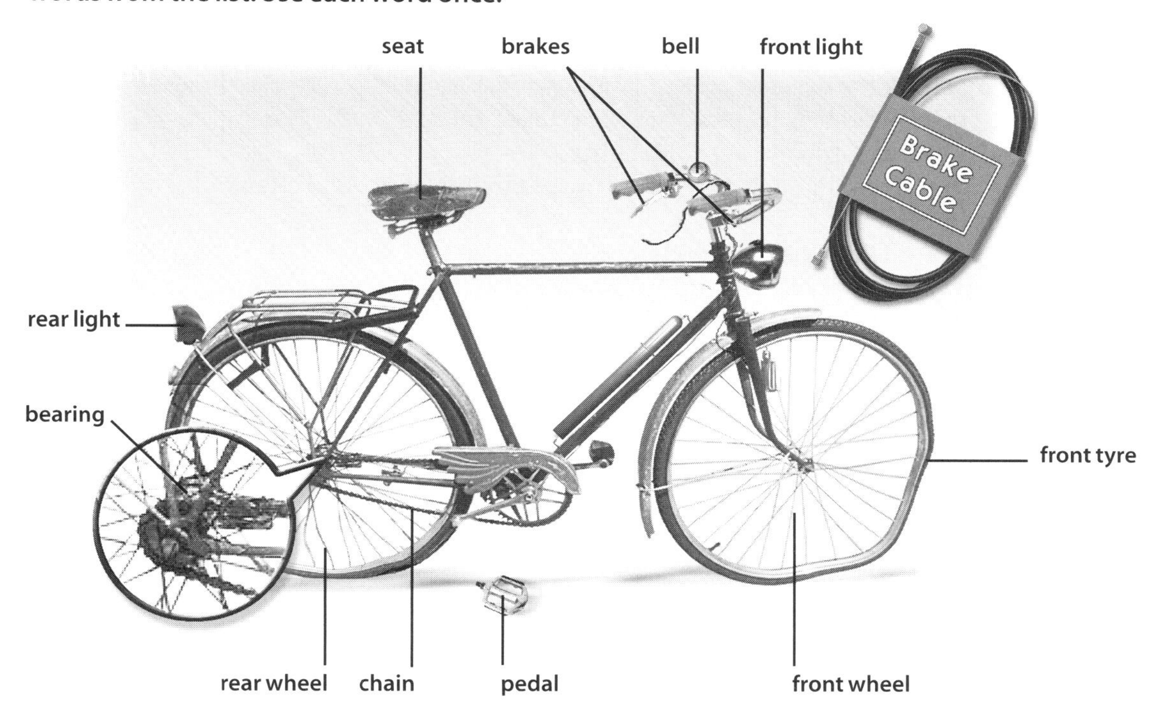

replacing	tightening	changing	pumping up
lubricating	straightening	lowering	oiling

Can you fix this bicycle by tomorrow? Here's a list of the things that need doing.

- *The brakes don't work. The cable is worn and it needs* 1 ...*replacing*...
- *The wheels are bent, so they need* 2 .... *And the flat tyres need* 3 ....
- *The seat is too high. It needs* 4 ....
- *Are the bulbs in the lights burnt out? If they are, they need* 5 ....
- *The chain is rusty. It needs* 6 .... *The wheel bearing needs* 7 .... *too, and the bell.*
- *I think some nuts and bolts are loose. They need* 8 ....

Thanks,
Larry Zimmerman

2 Larry forgot one thing. What other part needs fixing?

3 Read the conversation and complete the checklist.

A Mr Zimmerman's bicycle is ready.
B Good. Did anything need replacing?
A Yes, both the tyres.
B What about the wheels? Did you check them too?
A Yes, I did. I replaced the rear wheel.
B Did you replace the front wheel?
A No. I straightened it, but it didn't need replacing.
B Did you check the brakes?
A Yes. Both brake cables were worn so I replaced them.
B OK. Was there a problem with the lights?
A Yes, with the rear light. The bulb was burnt out. It just needed changing.
B So did you change it?
A Yes, I did. And I repaired the pedal too.
B And did you tighten all the nuts and bolts?
A Yes, and I oiled everything that needed lubricating.
B Good.
A The bell works fine now. But it'll need a new chain soon.
B Did you replace it?
A No, I didn't. I'll replace it next time.

		Action	Comments
1	☐	Check tyres	*Replaced front and rear tyres*
2	☐	Check wheels	
3	☐	Check brakes	
4	☐	Check lights	
5	☐	Check bell	
6	☐	Check pedals	
7	☐	Oil chain and bearings	
8	☐	Tighten nuts and bolts	

4 Did anything else need doing? What didn't he do? Look back to Mr Zimmerman's note and check.

5 Match the words on the left with the phrases on the right.

1 *Clean*
2 Delete
3 Download
4 Back up
5 Empty
6 Run
7 Update
8 Check

a your documents and files on a CD.
b the Recycle Bin.
c the Disk Cleanup and Defrag programs.
d any files you don't need.
e the hard drive for viruses.
f the drivers.
g *the screen, keyboard, and mouse.*
h any new service packs.

6 Complete the conversation with the words in the list.

did (x3)	didn't (x2)	checked	deleted	backed up
run	updating	need	download	

A I worked on your computer this morning. It's ready now.

B Great. What was the problem?

A I'm not sure, but it's OK now.

B [1] *Did* you [2]................. the Disk Cleanup and Defrag programs?

A Yes, I [3]................. .

B Did you [4]................. the new service packs?

A No, sorry, I [5]................. . Some of the programs are too big to download. You [6]................. a faster Internet connection.

B OK. What else did you do?

A Well, most of your drivers needed [7]................. . And I [8]................. two or three programs that you don't need.

B [9]................. you do a back-up?

A Yes, I [10]................. all your files. They're on these three CDs.

B Good ... Maybe the problem was a virus.

A No, I [11]................. your hard drive for viruses and it was OK. I cleaned everything, but I [12]................. empty the Recycle Bin.

B OK. Thanks a lot.

A You're welcome.

7 Write three more conversations. Follow the example.

need to measure this room / hold the tape measure?

A *Can you give me a hand?*
B *What's the problem? / What's up?*
A *I need to measure this room.*
B *Do you want me to hold the tape measure?*
A *Yes, please.*
B *No problem.*

1 can't see the connections inside the unit / hold the torch?

A *Are you busy?*
B ....
A ....
B ....
A ....
B ....

2 can't reach that button / press it for you?

A *Can you give me a hand?*
B ....
A ....
B ....
A ....
B ....

3 the pump isn't working / take a look?

A *Can you do me a favour?*
B ....
A ....
B ....
A ....
B ....

8 Match the things on the left with the problems on the right.

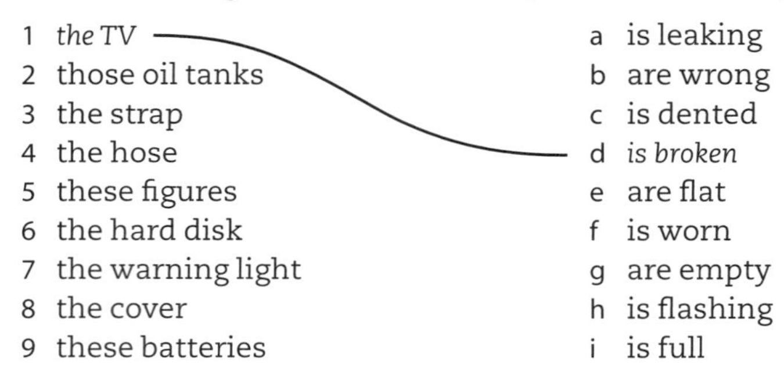

1 *the TV* — d
2 those oil tanks
3 the strap
4 the hose
5 these figures
6 the hard disk
7 the warning light
8 the cover
9 these batteries

a is leaking
b are wrong
c is dented
d *is broken*
e are flat
f is worn
g are empty
h is flashing
i is full

9 Read the email and put the sentences in the right order.

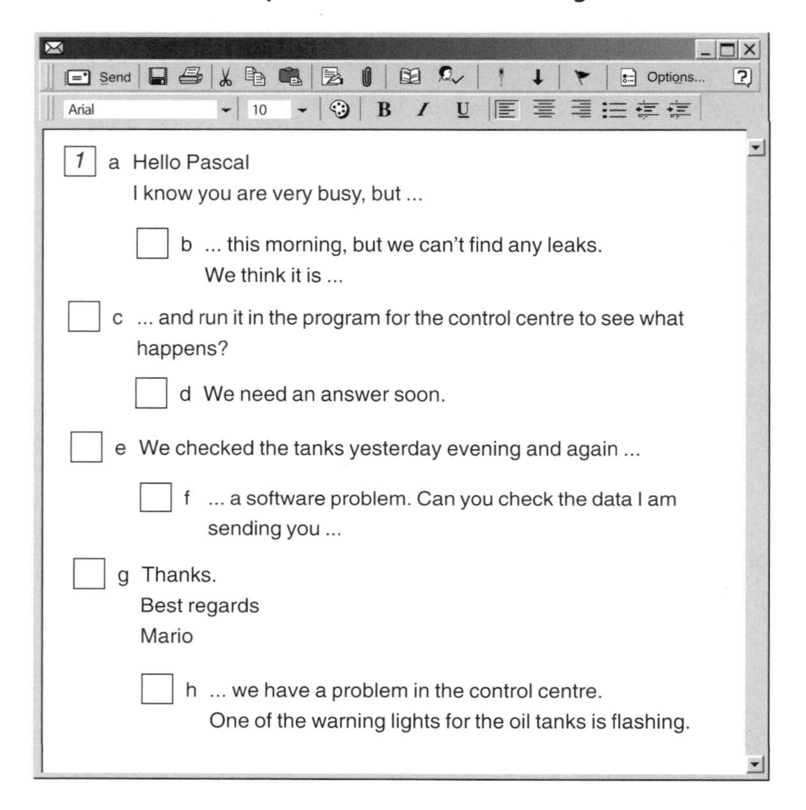

[*1*] a Hello Pascal
I know you are very busy, but ...

[ ] b ... this morning, but we can't find any leaks.
We think it is ...

[ ] c ... and run it in the program for the control centre to see what happens?

[ ] d We need an answer soon.

[ ] e We checked the tanks yesterday evening and again ...

[ ] f ... a software problem. Can you check the data I am sending you ...

[ ] g Thanks.
Best regards
Mario

[ ] h ... we have a problem in the control centre.
One of the warning lights for the oil tanks is flashing.

10 Read the answer to the email in 9. Complete the email with the words and phrases in the list.

What happens if
you need to
Did you check
Please
I checked
Do you want me to

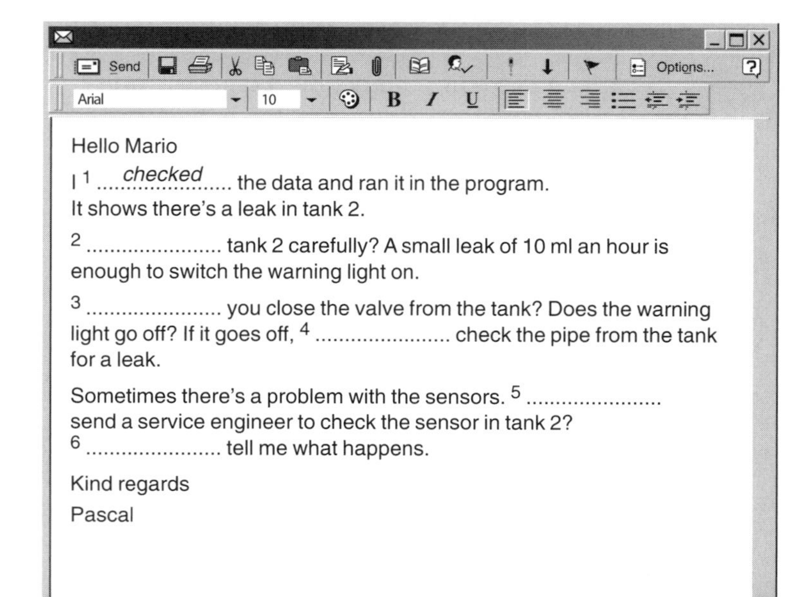

Hello Mario

I [1] *checked* the data and ran it in the program.
It shows there's a leak in tank 2.

[2] .... tank 2 carefully? A small leak of 10 ml an hour is enough to switch the warning light on.

[3] .... you close the valve from the tank? Does the warning light go off? If it goes off, [4] .... check the pipe from the tank for a leak.

Sometimes there's a problem with the sensors. [5] .... send a service engineer to check the sensor in tank 2?
[6] .... tell me what happens.

Kind regards
Pascal

Unit 20

1 Match the signs to the dialogues.

1 ..E..
A Don't touch any of the chemicals in the tanks here.
B Why not? Are they poisonous?
A No, but they can burn your skin.

2 .....
A Can we smoke in here?
B No, you mustn't smoke in here. There are flammable gases in those bottles.

3 .....
A What's that smell?
B I don't know, but I feel sick!
A Quick! We have to get out of here.

4 .....
A Is it a dangerous place to work?
B Yes, it is. You have to wear a hard hat and goggles.

5 .....
A Watch out! Don't put your hand on that machine.
B Why not?
A It's a twenty-tonne press!

6 .....
A It was in a cola bottle!
B Did you drink it?
A I drank a little.
B Well, you must see a doctor right away – I think it's old engine oil.

7 .....
A Before we go on the tour, you have to put on these special clothes and safety shoes.
B Is it dangerous in there?
A No, but it's better to be safe than sorry.
B That's true!

2 Give the safety instructions in a different way. Use the phrases in the list.

Never ... Don't ... Always ...

1 You mustn't put your hands near the blade.
Never put your hands near the blade.

2 You have to wear a hard hat in this area.
....

3 You mustn't remove the safety guard.
....

4 You mustn't lock the fire door.
....

5 You have to keep this area clean.
....

6 You must switch the machine off after you use it.
....

7 You mustn't touch the cable.
....

8 In case of fire, you must use the stairs.
....

3 Read the safety instructions for using a brushcutter. Complete the instructions with the words in the list.

allow	check	cut	wear	Fill up	Keep
Make sure	operate (×2)	protect	smoke	touch	Turn off

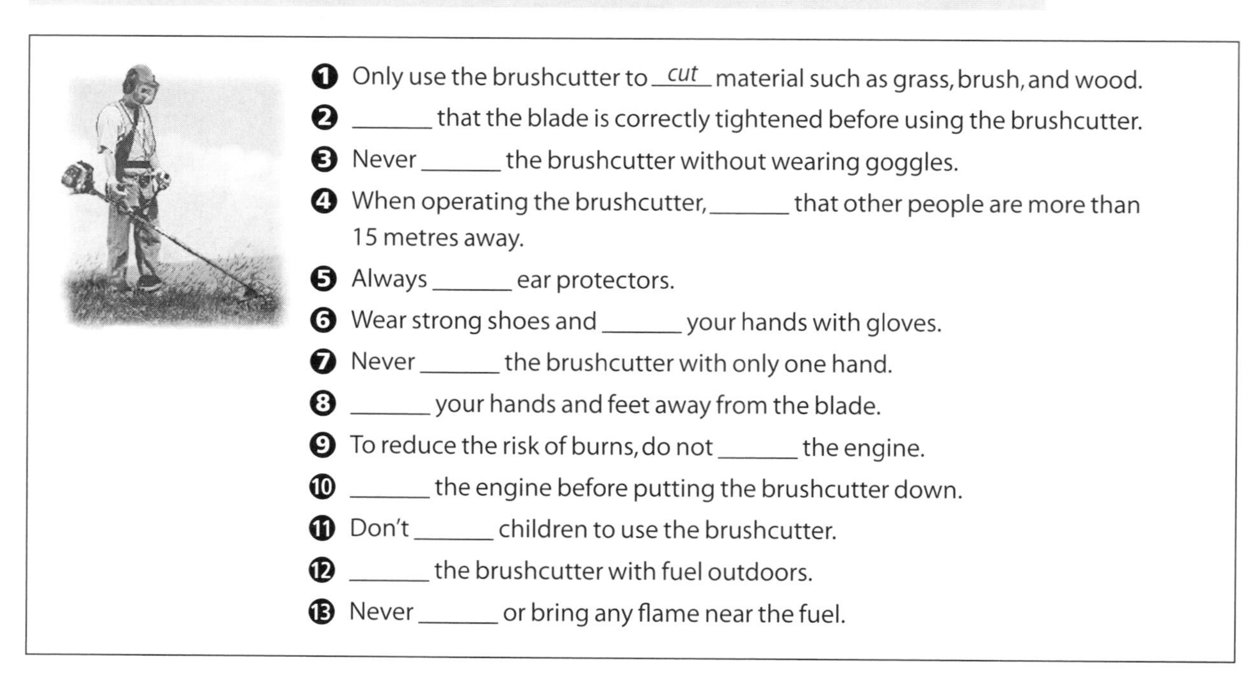

1. Only use the brushcutter to *cut* material such as grass, brush, and wood.
2. ______ that the blade is correctly tightened before using the brushcutter.
3. Never ______ the brushcutter without wearing goggles.
4. When operating the brushcutter, ______ that other people are more than 15 metres away.
5. Always ______ ear protectors.
6. Wear strong shoes and ______ your hands with gloves.
7. Never ______ the brushcutter with only one hand.
8. ______ your hands and feet away from the blade.
9. To reduce the risk of burns, do not ______ the engine.
10. ______ the engine before putting the brushcutter down.
11. Don't ______ children to use the brushcutter.
12. ______ the brushcutter with fuel outdoors.
13. Never ______ or bring any flame near the fuel.

4 Read the instructions in 3 again.

1 Decide if each instruction is something you *have to* do or something you *mustn't* do.
2 List the things you have to wear when you use a brushcutter.

5 Complete the dialogues with *will*, *'ll*, or *won't*.

1 A Be careful with those glass tubes.
B Why?
A They *'ll* break.

2 A Don't leave those pipes there.
B Why not?
A Someone .... trip over them.

3 A Write down the number.
B Why?
A You .... forget it.
B No, I ....

4 A Don't cut that cable.
B Why not?
A It .... be too short.

5 A Slow down.
B Why?
A We .... be late – there's an hour before your train leaves.

6 A Can you pass me the oil?
B Why?
A This bolt is too tight. It .... turn.

7 A Don't leave that outside.
B Why not?
A It .... go rusty.

6 Write the Past Simple form of these verbs. If you need help, look at the list of irregular verbs on page 117 of the Student's Book.

Present Simple	Past Simple	Present Simple	Past Simple
1 be (am / is)	*was*	11 lose	....
2 break	....	12 make	....
3 come	....	13 meet	....
4 do	....	14 pay	....
5 drink	....	15 run	....
6 drive	....	16 see	....
7 eat	....	17 sell	....
8 get	....	18 sit	....
9 give	....	19 take	....
10 go	....	20 wear	....

7 These sentences are all from accident reports.
Complete the sentences with the Past Simple form of some of the verbs in 6.

1 He _wore_ sports shoes to work, not safety shoes. He slipped on a wet floor and _broke_ his leg.

2 She ________ some engine oil that was in a cola bottle.

3 The truck ________ into the car park and hit two cars.

4 I lost my car keys, so I ________ the window. There ________ some broken glass on the driver's seat. I didn't see the glass. I ________ on the seat and cut myself.

5 She touched a live cable and ________ a 220-volt shock. After the accident, she ________ home. She ________ back to work the next day.

6 She ________ across the road to catch the bus. A bicycle hit her.

7 A security guard ________ smoke coming from the building at 11.30 p.m.

8 Someone ________ the safety cover off the machine. The accident happened the next day.

8 Read the story and change the verbs into the Past Simple.

That rattling noise!

Last year, a man (buy) 1 bought a new and very expensive car. The car (be) 2.................. perfect, except for one thing. There (be) 3.................. a rattling noise when he (go) 4.................. round corners, or (drive) 5.................. over bad roads. The man (take) 6.................. the car back to the garage where he (buy) 7.................. it. They (check) 8.................. everything, but the rattling noise (go) 9.................. on.

Finally, the man (tell) 10.................. the garage to look everywhere in the car to find the noise. The mechanics (remove) 11.................. everything – the engine, the gearbox, the seats ... everything! And inside one of the doors they (see) 12.................. the reason for the noise. There (be) 13.................. several nuts and bolts, (tie) 14.................. to a piece of string, with a note (attach) 15.................. . The note (say) 16.................. : 'I guess you finally (find) 17.................. the rattle'.

Unit 21

1 Read the text and label diagram A using the words in the list.

A Hydraulic elevator

tank piston pump car valve

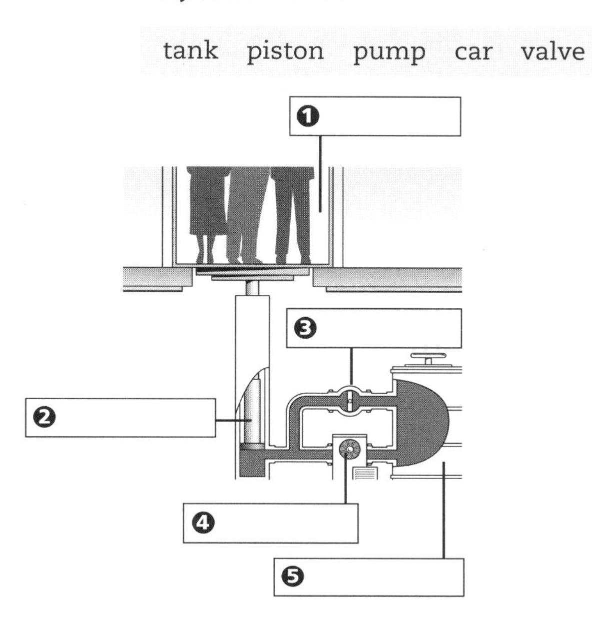

Hydraulic elevators

The car for transporting the passengers is raised and lowered by a hydraulic ram – an oil-driven piston inside a cylinder. The cylinder is connected to a fluid-pumping system. There are three parts to the system: a tank, a pump, and a valve.

The tank is at the end of the system. There is a pump between the the tank and the cylinder. It pushes the oil into the cylinder. The pressure inside the cylinder increases and pushes the piston up, raising the car.

When the car reaches the correct floor, the control system sends a signal to the electric motor in the pump. This switches the pump off. The car stays where it is.

To lower the car, the control unit sends a signal to the valve. The valve is located in a pipe above the pump. When the valve opens, the weight of the car pushes down on the hydraulic ram and the oil flows slowly back into the tank. To stop the car at a lower floor, the control system closes the valve again.

2 Read the text and number the parts in diagram B.

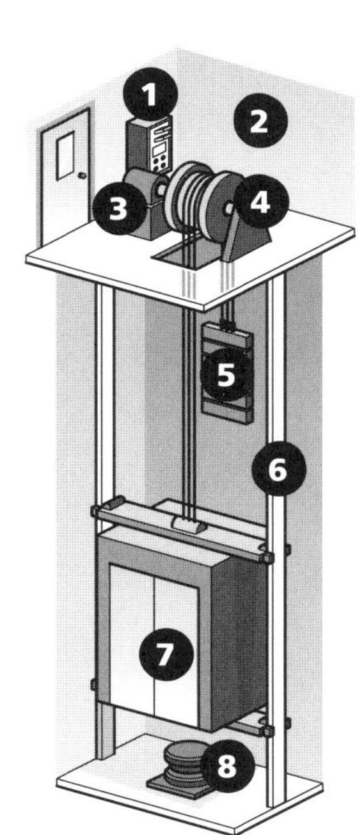

B Roped elevator

- .....1 control unit
- ..... counterweight
- ..... electric motor
- ..... elevator car
- ..... guide rails
- ..... winch
- ..... shock absorber
- ..... machine room

Roped elevators

The machine room is at the top of the elevator shaft. There is a control unit in the left-hand corner of the room. There is an electric motor between the control unit and the winch. It turns the winch and lifts the car up and down the elevator shaft.

The car is in the elevator shaft and is connected to three steel ropes. They run from the car, over the winch in the engine room to a counterweight at the back of the shaft.

The car and the counterweight ride on guide rails. The guide rails run along the sides of the shaft and at the back of the shaft. They stop the car and counterweight from hitting each other.

At the bottom of the shaft there is a shock absorber. It is there if something goes wrong and the car drops down the shaft.

3 Choose the correct verb to complete the sentences.

1 A piston <u>*pushes*</u> / *rotates* / *rolls* the car up the elevator shaft.
2 An electric motor *lowers* / *turns* / *pulls* the winch.
3 When the winch *blows* / *rolls* / *lowers* the counterweight, this raises the car.
4 The winch can *raise* / *lower* / *rotate* clockwise and anti-clockwise.
5 The car is *supported by* / *suspended from* / *pushed by* steel ropes.
6 The cylinder is *supported by* / *connected to* / *suspended from* a tank.
7 If the car *rotates* / *drops* / *burns* down the shaft, the shock absorber at the bottom of the shaft saves the passengers.

4 Use the words in the list to complete the phrases. Make as many combinations as you can.

cable *light* oil paper petrol rope steel string tape water wood

1 a beam of *light* ....
2 a piece of ....
3 a roll of ....
4 a tank of ....

5 What do *it* and *they* refer to in these sentences?

1 The sun shines onto the solar panels and they change the sunlight into electricity. *solar panels*
2 The cable runs to these sockets. It's about 10 metres long. ....
3 You need four AAA batteries for this torch, but they aren't expensive. ....
4 The cables are attached to a counterweight. They're made of steel. ....
5 The display is on the right-hand side of those buttons. It shows the temperature inside the engine. ....
6 The robot's arm has three fingers at the end. They can be programmed to pick up parts. ....

6 Space elevator

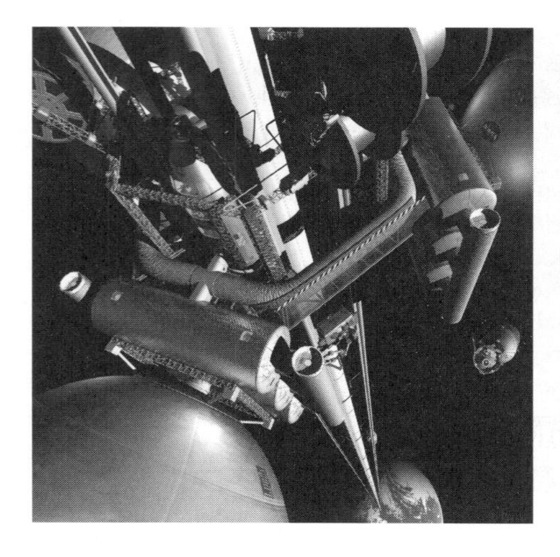

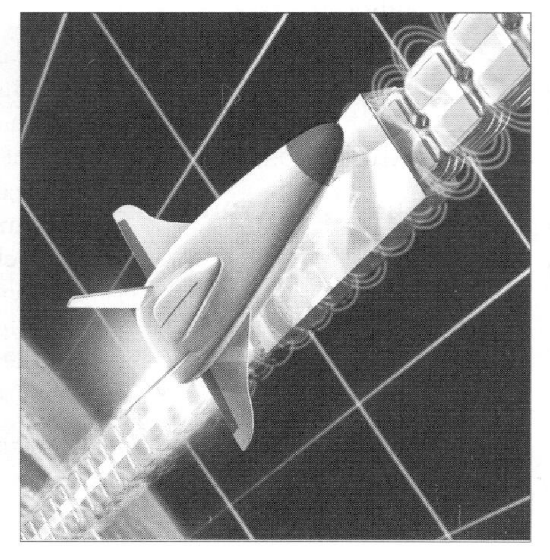

Read the dialogue. Complete the sentences with the words and phrases in the list.

A What's that?
B It's an idea for a space elevator.
A A space elevator? What's it for, exactly?
B It's for launching spacecraft, like the shuttle.
A How does it work?
B First we need to build a very tall tower – about 50 km high.
A 50 km!
B That's right. It needs to connect the cable to the earth.
A What's the cable for?
B The cable is attached to the earth. It runs to the top of the tower. From there it's connected to a weight. The weight is in orbit around the earth, above the equator.
A How long is the cable?
B It's about 144,000 kms long. It's made of carbon. It's lightweight and 100 times stronger than steel.
A I still don't really understand how it works.
B OK. The spacecraft travels up the tower, along the cable. We can use electricity to power it. This saves a lot of rocket fuel. At the top of the tower, the spacecraft rides up the cable towards the weight. At the end of the cable, the spacecraft has a speed of about 11 kilometres a second. That's fast enough to launch it into space.

saves   made of   long   connected to   *is for*   At the end of   attached to   needs to be

1 The space elevator ....*is for*.... launching spacecraft.
2 The tower is .... the earth.
3 The tower .... about 50 km high.
4 The cable is .... carbon.
5 The cable is .... the earth and a weight.
6 The carbon cable is about 144,000 km ....
7 .... the cable the spacecraft is moving at about 11 km/s.
8 This idea .... a lot of rocket fuel.

7 Match the pictures with the activities.

1 gardening ..*F*..
2 cycling .....
3 washing the car .....
4 cooking .....
5 going for long walks .....
6 playing football .....
7 travelling .....
8 reading .....
9 talking about work .....

8 **Now write some sentences about you. Start like this.**

I love ...

I like ...

I hate ...

Example *I love playing football.*

I hate washing the car.

9 **Match the sentences with the objects in the photos.**

A

B

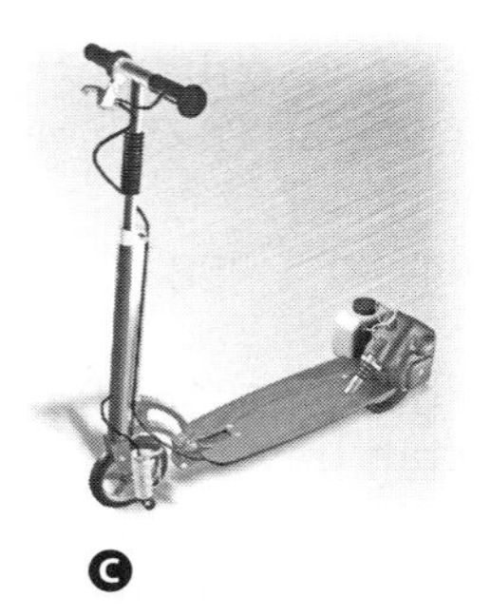

C

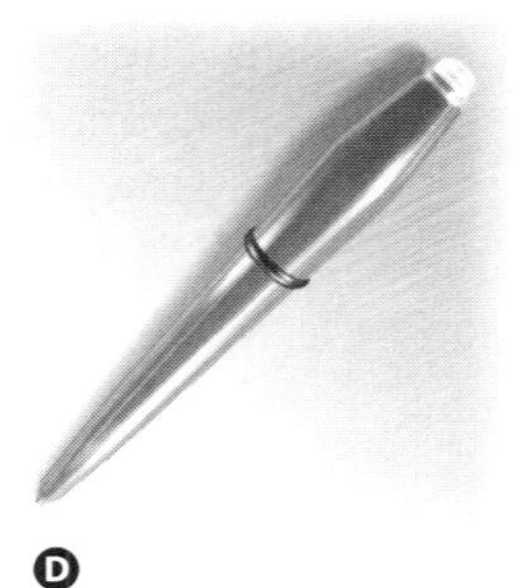

D

1 It's environmentally-friendly and uses the sun to produce heat. ..A...
2 It has a rear brake. .....
3 It has a built-in radio-alarm. .....
4 It can produce temperatures of between 200–450°C. .....
5 It shows the time on an LCD display. .....
6 It can write for three miles. .....
7 It can be used under water and works upside down. .....
8 It has a top speed of 22 km/h. .....
9 It's cylindrical. .....
10 An electric motor is connected to the rear wheel . .....
11 It can beam the time on to your bedroom wall. .....
12 It can't be used at night or inside a building. .....

Answer Key

Unit 1

1 2 meet 3 meet 4 too 5 Welcome 6 nice 7 Thank 8 speak 9 little 10 slowly

2 2 c 3 d 4 h 5 a 6 g 7 b 8 e

3 2 547 3 031 4 28/668 5 280 9345 6 5-602/5 7 6223 8 0081 9 437086 10 44/908-651

4 2 a 3 an 4 a 5 an 6 a 7 an 8 a **9** a 10 an

5 2 work 3 clock 4 alarm 5 torch 6 pencil 7 ticket 8 manual 9 machine 10 battery

6 2 pen 3 map 4 newspaper 5 umbrella 6 bag 7 key 8 alarm clock

7 2 e 3 a 4 b 5 h 6 d 7 f 8 c 9 j 10 i

8 **Conversation 2**

1 *Hello, I'm Paolo Castillo.*
4 Thank you.
3 It's nice to meet you, too. Welcome to Electronika.
2 Hello, I'm Dimitris Krassakis. Nice to meet you.

Conversation 3

2 Yes, just a little.
1 *Can you speak English?*
3 OK, no problem. I'll speak slowly.
4 Thanks.

Conversation 4

2 No, I'm sorry.
4 Me too. So please speak slowly.
1 *Can you speak French?*
3 I only speak a little English.
5 OK. No problem.

Unit 2

1 2 e 3 i 4 v 5 h 6 i 7 r 8 y 9 a

2 2 tea – b,c,d,e,g,p,t,v 3 my – i, y 4 pay – a,h,j,k 5 no – o 6 car – r

3

day /eɪ/	see /iː/	sell /e/	buy /aɪ/	go /əʊ/	you /uː/	are /ɑː/
a	b	f	i	o	q	r
h	c	l	y		u	
j	d	m			w	
k	e	n				
	g	s				
	p	x				
	t	z				
	v					

4 35 thirty-five 72 seventy-two 49 forty-nine 16 sixteen 28 twenty-eight 60 sixty 81 eighty-one 93 ninety-three 57 fifty-seven 27 twenty-seven 82 eighty-two 39 thirty-nine 13 thirteen

5 2 forty-nine 3 nineteen 4 eighty-nine 5 forty-eight 6 nineteen

6 2 what's your name 3 How do you spell that 4 my last name's Maier 5 I work for BMW 6 Yes, of course 7 Thank you 8 You're welcome

7 2 h 3 b 4 j 5 a 6 c 7 i 8 g 9 f 10 d

8

X Y W A S H E R S Z
S J X Z Y O X Q Y B
Q P J X Z O Y B J A
X Q R J Z K Q O Z T
Z C L I P X J L X T
J X Z Q N Y X T Z E
W O Q Z X G R X J R
C L A M P X S Y Q I
H T W V Y Z Q Z X E
J D X R O P E X V S

9 1 Chi & Co. Ltd.
2 East African Technologies Ltd.
3 Equipamentos Indústrias Ltda.
4 service technician, ABR Danmark
5 Singapore, EnTech Corporation, software engineer
6 Paola Belisario

Unit 3

1 1 burger 2 hot dog 3 Diet-Cola 4 milkshake 5 coffee 6 French fries 7 doughnut 8 cheeseburger 9 sandwich

2 2 Can I help you 3 Large, medium, or small 4 please 5 a large 6 And something to drink 7 Can I have 8 Thanks 9 How much is it 10 Have a nice day

3 2 a 3 d 4 g 5 b 6 c 7 f

4 1 Can I use your mobile phone?
Do you want to make a call?
No, I want to send a text message.
2 Can I have the order form?
Do you want to change the order?
No, I want to check the quantity.
3 Can I look at your manual?
Do you want to read the instructions?
No, I want to check a serial number.

5 3, 5, 6 correct 4 75-watt 7 six-volt 8 five-litre

6 2 1 yd 4 1 gal 6 1 cm 8 1 g
3 1 m 5 1 lb 7 1 ft

7
Length
metres
centimetres
millimetres
yards
feet
inches

Volume
litres
millilitres
gallons

Weight
tonnes
kilograms
grams
milligrams
ounces

8
2 want
3 no good
4 Do you want
5 this bolt
6 Is it
7 it is
8 I need a
9 Do you have
10 How much is it

9 10 mm spring
12-volt battery
60 mm bolt
900 mm cable

Unit 4

1
2 Can I
3 please
4 I'm afraid
5 give
6 can't
7 take
8 Just
9 need
10 It's
11 spell
12 ask
13 call
14 that's
15 correct
16 with
17 You're welcome

2 Message 3

3 2 f 3 b 4 h 5 a 6 i 7 c 8 e 9 g

4
3 need
4 have
5 Do you
6 do you
7 have
8 don't have
9 you have
10 do you
11 have
12 don't have

5
1 wall socket
2 extension lead
3 business card
4 serial number
5 installation disk
6 telephone number
7 email address

6 *See diagram on page 80.*

7 2 c 4 a 6 b 8 c 10 b
3 c 5 a 7 a 9 a

Unit 5

1 1 H 3 D 5 A 7 C
2 E 4 F 6 B 8 G

2
2 There are
3 There are
4 There's
5 There's
6 There are
7 There's
8 There's

3 2 8 4 2 6 #
3 7 5 *

4 *See diagram on page 80.*

5
2 top
3 correct
4 small
5 Press
6 switches
7 centre
8 display
9 button
10 right
11 lock
12 replace

6
2 close
3 remove
4 turn on
5 pull
6 turn clockwise
7 plug in

7
2 Read
3 Turn
4 Pull
5 Unlock
6 Replace
7 Press
8 Remove

8
2 Switch on, open, read
3 close, switch off, lock
4 Switch off, remove, replace
5 push, turn, turn off

9
2 cars
3 businesses
4 batteries
5 people (irregular)
6 days
7 bulbs
8 torches
9 cities
10 months
11 items
12 messages

10 *See diagram on page 80.*

11
2 Yes, there are.
3 No, there isn't.
4 No, there isn't.
5 Yes, there are.
6 No, there aren't.
7 No, there isn't.
8 Yes, there are.

Unit 6

1 2 K 4 F 6 B 8 G 10 I 12 D
3 J 5 H 7 C 9 A 11 L

2
2 Does it have
3 No, it doesn't
4 Does it have
5 No, it doesn't
6 Does it have
7 No, it doesn't
8 Do you want

3 Extras: CD player, automatic transmission, head air bags, navigation system, central locking

4 2 T 3 F 4 T 5 F 6 T 7 T 8 F

5
3 It's made of glass.
4 It's made of wood.
5 It's made of plastic.
6 They're made of paper.
7 It's made of aluminium.
8 They're made of leather.
9 They're made of steel.
10 It's made of paper.

6
2 No, it doesn't.
3 No, it doesn't.
4 Yes, it does.
5 Yes, it does.
6 No, it doesn't.
7 Yes, it does.
8 Yes, it does.

7
2 sockets
3 fuel tank
4 buttons
5 words
6 drive
7 facilities
8 é
9 numbers
10 bulb

8
2 semi-circle, semi-circular
3 oval, oval
4 cylinder, cylindrical
5 triangle, triangular
6 sphere, spherical
7 cube, cubic
8 square, square

9
2 It's rectangular.
3 It's square.
4 It's triangular.
5 It's cylindrical.
6 It's semi-circular.
7 It's spherical.
8 It's oval.
9 It's cubic.

Unit 7

1
1 face
2 nose
3 mouth
4 front leg
5 foot
6 head
7 neck
8 back
9 back leg

2 2 T 4 F 6 F 8 T 10 F
3 T 5 T 7 F 9 T

3 2 E 3 G 4 C 5 B 6 I 7 H 8 F 9 D

4
2 climb
3 push
4 raise
5 move
6 carry
7 adjust
8 bend
9 turn
10 straighten

5
2 chair
3 clock
4 cog
5 plane
6 crane
7 CD drive
8 car

6 1 D 2 F 3 A 4 E 5 C 6 B

7
2 wide
3 long
4 heavy
5 much
6 long
7 long
8 high

8
2 clean
3 small
4 can
5 downstairs
6 lower
7 backwards
8 turn left
9 straighten
10 push
11 turn off
12 open

Unit 8

1
1 install
2 check
3 replace
4 replace
5 install

2
2 it
3 them
4 them
5 them
6 it
7 them
8 it

3
2 order
3 straighten
4 check
5 measure
6 replace
7 clean
8 delete

4 2 one 6 one 10 some
3 some 7 a 11 a
4 any 8 one 12 some
5 a 9 a 13 any

5 1 c 3 b 5 e 7 h 9 i
2 a 4 f 6 j 8 d 10 g

6 2 G 5 H 8 N 11 D 14 J
3 E 6 L 9 B 12 P 15 C
4 K 7 A 10 M 13 F

7 2 key
3 magnifying glass
4 fork-lift truck
5 funnel
6 pair of scissors
7 crate / box
8 thermometer
9 spanner / wrench
10 pair of goggles
11 scales
12 can of oil
13 switch
14 torch
15 tape measure

8 1 b 2 e 3 g 4 f 5 a 6 c 7 d

Unit 9

1 2 I 5 A 8 F 11 E 14 D
3 C 6 K 9 O 12 M 15 N
4 H 7 J 10 G 13 B

2 Don't: 2 5 7 9 10

3 2 black 6 brown
3 blue 7 yellow
4 white 8 grey
5 green

4 3 should 7 should
4 shouldn't 8 shouldn't
5 shouldn't 9 should
6 should 10 shouldn't

5 2 Are 5 Is 8 Do
3 Is 6 Do 9 Does
4 Does 7 Are 10 Is

6 2 Yes, she is. 7 Yes, I do.
3 Yes, you do. 8 No, she doesn't.
4 No, I'm not. 9 Yes, I am.
5 Yes, he does. 10 Yes, it does.
6 No, she isn't.

7 2 Do you work for an electronics company?
3 Is your company in the aerospace industry?
4 Does your company produce mobile phones?
5 Do you need to have a visitor's pass?
6 Is there a photocopier on the first floor?
7 Are you Mr Linberg from Nordea in Oslo?

Unit 10

1 2 L 4 I 6 G 8 D 10 E 12 C
3 J 5 H 7 K 9 B 11 F

2 2 on 6 on, under 10 on
3 between 7 in 11 in front of
4 behind 8 between 12 next to
5 next to 9 under

3 2 No, it isn't. 5 No, there isn't.
3 No, it isn't. 6 Yes, it is.
4 No, it isn't

4 2 D 4 F 6 A 8 B
3 E 5 H 7 C

5 2 It's a quarter past eight.
3 It's four thirty.
4 It's twenty-five past six.
5 It's three thirty.
6 It's three.
7 It's a quarter to twelve.
8 It's one fifty.

6 2 7.35 4 3.35 6 7.10
3 10.10 5 12.45

7 1 this, that 3 These, Those
2 these, that 4 these, that

Unit 11

1 2 not 5 don't 8 need to
3 wrong 6 can't 9 have
4 wrong 7 don't 10 can't

2 2 E 3 B 4 A 5 F 6 D

3 2 it's too far.
3 they're too small.
4 it's too old.
5 it's too dangerous.
6 they're too heavy.
7 it's too small.
8 they're too difficult.

4 2 A 3 C 4 F 5 H 6 G 7 B 8 D

5 Drive into the company, meet the Chief Engineer, bring a video camera and laptop, go on a tour of the labs.

6 2 have to 5 can't 7 have to
3 have to 6 have to 8 can't
4 can't

7 2 g 3 d 4 a 5 e 6 b 7 f

8 2 T 3 F 4 F 5 T 6 T 7 F 8 T

Unit 12

1 2 P 6 M 10 C 14 T 18 S
3 K 7 D 11 O 15 I 19 F
4 A 8 B 12 E 16 J 20 G
5 N 9 R 13 H 17 L

2 2 aren't using
3 are they doing
4 's removing
5 Is he repairing
6 's installing
7 are you doing
8 'm testing
9 are repairing
10 're soldering
11 are they wearing

3 2 I'm connecting the printer to the computer.
3 Be careful! You're standing on the cables!
4 She's reading the manual in English.
5 We're building a new workshop.
6 They're testing the machine now. It's working well.

4 2 Lorenzo's not / isn't operating the machine today.
3 We're not / We aren't replacing the batteries.
4 She's not / She isn't doing the tests for Mr Nikolaev.
5 They're not / They aren't using the electric drill. It's not / It isn't working.
6 I'm not checking the connections at the moment.

5 2 Is she adjusting it?
3 What are you soldering?
4 Why are they changing the computer?
5 Are the batteries leaking?
6 Why is he removing the cover?

6 **Countable:**
tape measure, pair of pliers, screwdriver, boxes, wrenches, spirit level, cables, saws, electric drill, pair of goggles, manuals, hammer, bottles, printer, keyboard.

Uncountable:
petrol, paint, paper, gas, string.

7 1 any, a
2 some, a, some
3 any, some
4 some, a
5 any, some, an, a, any

8 2 e 3 d 4 b 5 a 6 f

9 2 bolt 4 cement 6 screw
3 nail 5 glue

10 2 cement 4 glue 6 bolt
3 staple 5 nail

Unit 13

1 2 f 4 a 6 d 8 g 10 e
3 i 5 b 7 j 9 h

2 2 sunglasses 6 mirror
3 map 7 hat
4 knife 8 lighter
5 whistle

3 Students' own answers.

4 Conversation 1
correct order: a e c d b

Conversation 2
correct order: d c f b e a

Conversation 3
correct order: c e a f b d

5 1 rewind 3 volume 5 power
2 play 4 LCD screen

6 2 I 3 G 4 F 5 D 6 H 7 C 8 B 9 E

7 2 decrease 6 rewind
3 unplug 7 put down
4 pull 8 start / begin
5 close / shut

8 2 button 7 calls
3 for 8 how
4 enter 9 button
5 Use 10 easy
6 What

Unit 14

1 2 f 3 h 4 b 5 g 6 a 7 c 8 e

2 2 were 6 was
3 was 7 were, was
4 was, were 8 were
5 was, were

3 (possible answers)
1 We repaired the power cord.
What was wrong with it?
It was worn.
2 She replaced the batteries.
What was wrong with them?
They were leaking.
3 I changed the fuses.
What was wrong with them?
They were burnt out.
4 I changed the tyre on the truck.
What was wrong with it?
It was flat.
5 We painted the pipes.
What was wrong with them?
They were rusty.
6 I fixed the handle on your bag.
What was wrong with it?
It was broken.

4 1 F 2 F 3 T 4 T 5 F 6 F 7 T 8 F

5 2 were 6 was
3 happened 7 contacted
4 used 8 arrived
5 started 9 was

6 2 at 4 on 6 at 8 on 10 in 12 at
3 at 5 on 7 in 9 on 11 on

7 3 tested •• 7 signed •
4 used • 8 asked •
5 counted •• 9 bolted ••
6 started •• 10 fixed •

8 2 the eighth of December, August twelfth
3 the sixth of March, June third
4 the fifth of September, May ninth
5 the eleventh of February, November second
6 the tenth of April, October fourth

9 2 began/started 11 collapsed
3 began/started 12 closed
4 stopped 13 signed
5 finished 14 installed
6 increased 15 removed
7 decided 16 corrected
8 installed 17 completed
9 used 18 visited
10 added

Unit 15

1 1 the USA (America) 13 watching baseball
2 lab assistant 14 playing golf
3 Elks Run 15 Cincinnati Reds
4 Chinese 16 mechanical engineer
5 Bergen 17 English
6 two 18 French
7 four 19 skiing
8 Norwegian 20 travelling
9 German 21 Peder
10 sailing 22 Ragnhild
11 walking
12 English

2 3 correct
4 How many languages do you speak?
5 Where is Kai from?
6 correct
7 Where do you live?
8 What sports do you like?
9 correct
10 How many children do you have?

3 2 Where does she live?
3 What does she do? / What's her job?
4 How many languages does / can she speak?
5 How many children does she have?
6 What are their names?
7 What are her hobbies?
8 Does she play golf?

4 Students' own questions.

5 2 g 4 h 6 d 8 j 10 c
3 i 5 b 7 a 9 e

6 2 three-quarters 5 a quarter
3 half 6 a fifth
4 two-thirds 7 half

7 2 F 3 T 4 T 5 T 6 F 7 F 8 F

8 2 b 4 b 6 c 8 c 10 b 12 b
3 a 5 b 7 a 9 a 11 c

Unit 16

1 2 Can you 5 Can I 7 Can I
3 Can I 6 Can you 8 Can I
4 Can you

2 2 Can I use the phone?
3 Can you turn off the air-conditioning?
4 Can I put my tools in the car / your car?
5 Can I use tape to insulate this wire?
6 Can you check these calculations?
7 Can I leave my bag at reception?
8 Can you finish the job today?

3 2 and 5 mean No.

4 2 h 3 a 4 g 5 d 6 c 7 b 8 e

5 (possible answers)
1 Can you lend me your mobile phone?
I'm afraid it's not charged.
OK, it doesn't matter.
2 Can I look at your manual?
Sure, go ahead. It's on the shelf.
Thanks.
3 Can I borrow your English dictionary?
I'm afraid Stefan is using it.
OK, it doesn't matter.
4 Can I borrow your car this afternoon?
I'm afraid there's no petrol in the tank.
OK, it doesn't matter.
5 Can I use the photocopier?
Sure, go ahead. You need a code – it's 3039 + Enter.
Thanks.
6 Can you give me some change for a coffee?
Yes, of course. Here's two euros.
Thanks.

6 1 500 3 1,000 5 700
2 850 4 880

7 2 take 7 It takes
3 takes 8 how long does it take
4 how long 9 about
5 pages 10 pages
6 What 11 How long

8 2 h 4 k 6 c 8 b 10 g
3 i 5 a 7 j 9 d 11 e

9 2 C 3 C 4 U 5 C 6 U 7 U 8 C

10 2 much 5 many 7 many
3 many 6 much 8 many
4 much

11 2 ✗ 3 ✓ 4 ✓ 5 ✗ 6 ✗

Unit 17

1 2 F 3 D 4 C 5 G 6 E 7 B 8 H

2 2 up 5 over 7 round
3 along 6 under 8 through
4 down

3 *See diagram on page 80.*

4 2 going 6 Go 9 going
3 come 7 go 10 come
4 go 8 go
5 come

5 2 taxi 5 station 7 airport
3 airport 6 station 8 taxi
4 airport

6 1 How many bags do you have?
2 Did you pack them yourself?
3 Have a good flight.
4 A ticket to Chicago, please.
5 When does the train leave?
6 At five to eight.
7 How much is that?
8 And I need a receipt.
9 Do you need a hand with your bags?

7 Silent letters: 2 u 3 k 4 w 5 u 6 h 7 w 8 c 9 d

8 1 Health and Safety Building
2 Final Inspection Building
3 Lab 2
4 Research & Development / R&D Building
5 Administration Building

Unit 18

1 a 9 c 1 e 2 g 3 i 4
b 10 d 6 f 5 h 7 j 8

2 2 j 4 e 6 d 8 c 10 i
3 a 5 g 7 h 9 b

3 2 loosen 9 connect
3 Remove/Take off 10 fix
4 Position 11 tighten
5 measure 12 check
6 cut 13 Replace
7 Remove/Take off 14 switch … on
8 Locate

4 2 Turn them on. 7 Key them in.
3 Fill it in. 8 Switch it off.
4 Hold them down. 9 Take it off.
5 Slow it down. 10 Put them on.
6 Take them off.

5 3 by pressing 7 to switch on
4 to set 8 by turning
5 to open 9 to increase
6 by removing 10 by breaking

6 2 burns 6 bursts
3 explodes 7 contracts
4 melts 8 rusts
5 evaporates

7 2 place 6 pressure
3 pump 7 press
4 lowers 8 opens
5 removes

8 1 b 2 a 3 b

Unit 19

1 2 straightening 6 oiling
3 pumping up 7 lubricating
4 lowering 8 tightening
5 changing

2 The pedal needs repairing.

3 2 Replaced rear wheel, straightened front wheel
3 Replaced both brake cables
4 Changed rear light bulb
5 Checked bell
6 Repaired broken pedal
7 Oiled chain and bearings
8 Tightened nuts and bolts

5 2 d 3 h 4 a 5 b 6 c 7 f 8 e

6 2 run 8 deleted
3 did 9 Did
4 download 10 backed up
5 didn't 11 checked
6 need 12 didn't
7 updating

7 (possible answers)

1 A Are you busy?
B What's up?
A I can't see the connections inside the unit.
B Do you want me to hold the torch?
A Yes, please.
B No problem.

2 A Can you give me a hand?
B What's the problem?
A I can't reach that button.
B Do you want me to press it for you?
A Yes, please.
B No problem.

3 A Can you do me a favour?
B What's up?
A The pump isn't working.
B Do you want me to take a look?
A Yes, please.
B No problem.

8 2 g 3 f 4 a 5 b 6 i 7 h 8 c 9 e

9 Correct order: a h e b f c d g

10 2 Did you check
3 What happens if
4 you need to
5 Do you want me to
6 Please

Unit 20

1 2 C 3 F 4 D 5 G 6 B 7 A

2 2 Always wear hard hats in this area.
3 Don't / Never remove the safety guard.
4 Don't / Never lock the fire door.
5 Always keep this area clean.
6 Always switch the machine off after you use it.
7 Don't / Never touch the cable.
8 In case of fire, always use the stairs.

3 2 Make sure 8 Keep
3 operate 9 touch
4 check 10 Turn off
5 wear 11 allow
6 protect 12 Fill up
7 operate 13 smoke

4 1 **You have to –**
instructions 1 2 4 5 6 8 10 12
You mustn't –
instructions 3 7 9 11 13

2 goggles, ear protectors, shoes, gloves

5 2 will 4 'll / will 6 won't
3 'll, won't 5 'll 7 'll / will

6 2 broke 9 gave 16 saw
3 came 10 went 17 sold
4 did 11 lost 18 sat
5 drank 12 made 19 took
6 drove 13 met 20 wore
7 ate 14 paid
8 got 15 ran

7 2 drank
3 drove
4 broke, was, sat
5 got, went, came / went
6 ran
7 saw
8 took

8 2 was 8 checked 14 tied
3 was 9 went 15 attached
4 went 10 told 16 said
5 drove 11 removed 17 found
6 took 12 saw
7 bought 13 were

Unit 21

1 1 car 4 pump
2 piston 5 tank
3 valve

2 2 machine room 6 guide rails
3 electric motor 7 elevator car
4 winch 8 shock absorber
5 counterweight

3 2 turns 5 suspended from
3 lowers 6 connected to
4 rotate 7 drops

4 2 cable, paper, rope, steel, string, tape, wood
3 cable, paper, tape
4 oil, petrol, water

5 2 cable 5 display
3 batteries 6 fingers
4 cables

6 2 connected to 6 long
3 needs to be 7 At the end of
4 made of 8 saves
5 attached to

7 2 H 3 A 4 G 5 C 6 B 7 I 8 E 9 D

8 Students' own answers.

9 2, 8, 10 – c
3, 5, 11 – b
4, 12 – a
6, 7, 9 – d

Diagrams

Unit 4 – Exercise 6

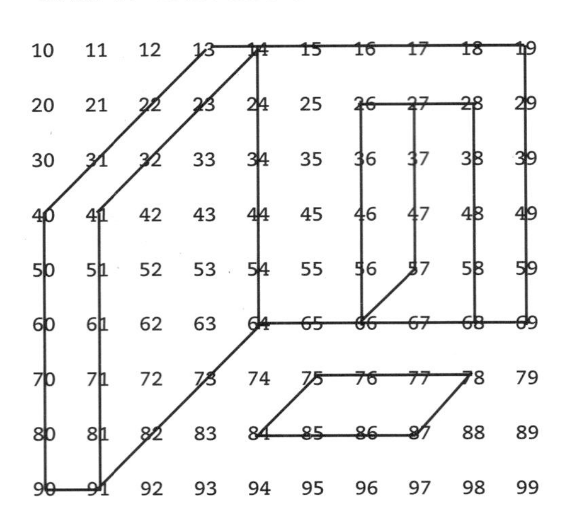

Unit 5 – Exercise 4

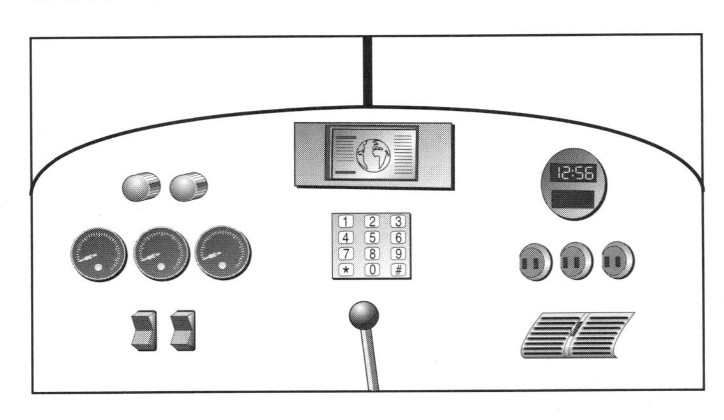

Unit 5 – Exercise 10

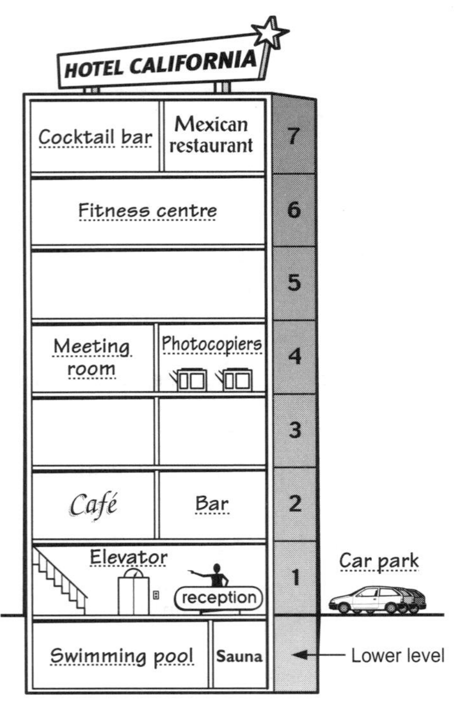

Unit 17 – Exercise 3

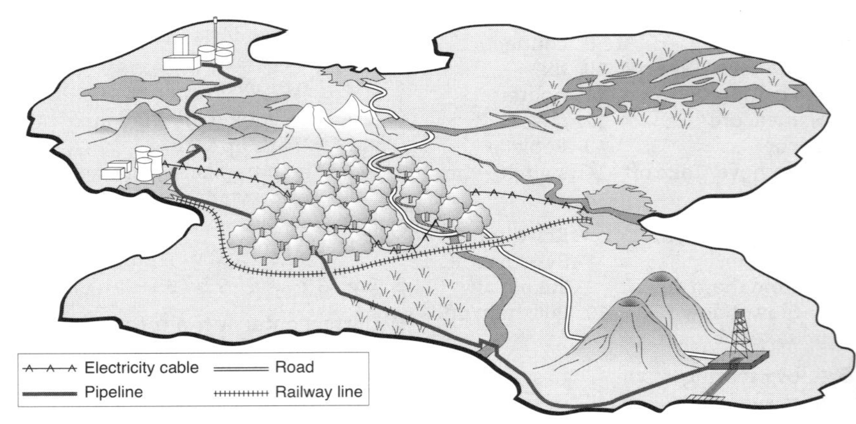